Religious Experience Reconsidered

Religious Experience Reconsidered

A BUILDING-BLOCK APPROACH TO THE STUDY OF RELIGION AND OTHER SPECIAL THINGS

Ann Taves

PRINCETON UNIVERSITY PRESS

PRINCETON AND OXFORD

Second printing, and first paperback printing, 2011
Paperback ISBN 978-0-691-14088-9

THE LIBRARY OF CONGRESS HAS CATALOGED THE CLOTH EDITION
OF THIS BOOK AS FOLLOWS

Taves, Ann, 1952–
    Religious experience reconsidered : a building-block approach to the
study of religion and other special things / Ann Taves.
        p.   cm.
    Includes bibliographical references and index.
    ISBN 978-0-691-14087-2 (alk. paper)
    1. Experience (Religion). 2. Meaning (Philosophy)—Religious aspects.
I. Title.
    BL53.T39 2009
    204'.2—dc22   2009006059

British Library Cataloging-in-Publication Data is available

This book has been composed in Sabon

Printed on acid-free paper. ∞

Printed in the United States of America

10 9 8 7 6 5 4 3 2

To Ray, with love

Contents

Illustrations and Tables

Acknowledgments

Although I have been preoccupied with the problem of religious experience for some time, the decision to write this particular book emerged in the wake of the Evolution of Religion Conference held in Hawaii in January 2007, where I was asked to give one of the plenary addresses. I am grateful to Joseph Bulbulia, Armin Geertz, and the other members of the organizing committee for providing an occasion for articulating my thoughts about studying "religious experience." I am also grateful to colleagues who provided important feedback on early drafts of chapters, including Tom Tweed and Ilkka Pyysiäinen on chapter 1, Robert Sharf on chapter 2, and Wayne Proudfoot and William Barnard on chapter 3. During Winter Quarter 2008, the students in my doctoral seminar—Robert Borneman, Jared Lindahl, Andrew Mansfield, Andrea Neuhoff, Albert Silva, and Kristy Slominski—hammered away at the first draft of the manuscript, especially the first chapter. Their feedback was invaluable and much of it has been incorporated in the final revision. In the Spring Quarter, Todd Foose and Brian Zeiden, my teaching assistants for an undergraduate course in psychology and religion, went over much of the material with me in another format. I am grateful to Melinda Pitarre for assistance in the coding of the personal narratives in chapter 3 and with the bibliography. Bill Christian, visiting professor at the University of California, Santa Barbara, during Spring Quarter, enthusiastically plied me with examples of "singularization" and "special things," and a conference on Religious Ritual, Cognition, and Culture at the University of Aarhus in May 2008 provided the occasion for further testing and revision of these ideas.

Special thanks are due to Fred Appel, my editor at Princeton University Press, and my husband, Ray Paloutzian, both of whom have read, edited, and commented on numerous drafts of every chapter. Fred, as the Press's senior editor for religion, music, and anthropology, made sure that the book was intelligible to humanists, while Ray, as a psychologist and a journal editor, did the same for the scientists. As an editorial tag team that saw eye to eye on questions of style, their concerted efforts made this a vastly more readable book. Several colleagues also read the whole manuscript and provided detailed feedback, including Catherine L. Albanese, my department chair, and two readers for Princeton University Press, one of whom was Tanya Luhrmann. Their suggestions, which I have done my best to incorporate in the final draft, polished it yet further. I would also like to thank my production editor, Heath Renfroe, and copyeditor,

Jon Munk, for the skill and care with which they moved the manuscript through production.

Since Ray and I met just as I was starting to write and married days after the first draft was completed, I am grateful to him for much more than reading and editing. As my companion and conversation partner in all aspects of life, I dedicate this book to him.

Preface

For reasons of temperament and training, I find it natural and exciting to make forays across what many scholars see as an unbridgeable divide between the humanities and the natural sciences. I must admit to a certain impatience with those of my fellow humanists who police these boundaries and caution against serious engagement with the natural sciences. In my view, it is better to construct rough and ready bridges than to wait for the construction of a perfect bridge that will stand for all time. This book is devoted to building some usable, albeit imperfect, bridges linking the study of experience in religious studies, the social-psychological study of the mind, and neuroscientific study of the brain.

I have written this book primarily for humanists and humanistically oriented social scientists who study religion using historical and ethnographic methods. My hope is that the conceptual tools provided here will embolden these readers to make greater use of scientific research that is illuminating the complex ways in which the brain-mind is both shaped by and shapes socio-cultural processes. I also hope that this book will be useful to experimentalists who study religion—to help them consider ways in which the resources of the humanities might enhance their experimental research designs or provide new contexts for testing hypotheses.

The focus of the book is on experiences deemed religious (and, by extension, other things considered special) rather than "religious experience." This shift in terminology signals my interest in exploring the processes whereby experiences come to be understood as religious at multiple levels, from the intrapersonal to intergroup. To understand these processes, I argue that we need to work comparatively, but that we cannot limit our comparisons to "religious things," as if "religious things" or "religious experiences" comprised a fixed and stable set. Rather, much as scientists compare experimental and control groups, we need to compare things that people consider religious with similar things that they do not. The phrase "experiences deemed religious" is contentious, as is each of the individual words "experience," "deemed," and "religious." A chapter is devoted to each word, starting with "religion," and followed by "experience" and then "explanation," which takes up "deeming." The fourth chapter—devoted to comparison—discusses how we might best set up comparisons between experiences that are sometimes considered religious and sometimes not.

Scholars of religion regularly raise certain objections to the approach I am advocating. First, they suggest that the subject matter is passé in an era that has abandoned experience for discourse *about* experience. Second, they worry that an approach that compares religious and nonreligious things will wind up being reductionistic—that is, it will "reduce" religion to something else. And, third, they offer critiques of scientific methods and claims drawn from science studies. While I do not deny the many legitimate concerns humanists have raised relative to scientific methods and claims, I do not think these concerns should stop us from engaging with research on the other side of the academic divide.

The book addresses the subject of religious experience directly and the problems of reductionism and humanistic fears of the sciences indirectly and by example. The orientation of the book is practical more than philosophical. In the process of identifying methods that will allow us to cross back and forth across this humanistic/scientific divide more easily and responsibly, I draw from work in religious studies, anthropology, history, philosophy of science, psychology, and neuroscience. In doing so, I sidestep contentious issues where possible, privileging method over theory and philosophy in the interests of actually crossing the divide, while alerting readers to the unresolved philosophical and theoretical issues in the notes.[1] The book is not intended to address all the thorny issues surrounding "religious experience" but is designed to alert researchers to some of the most hotly contested issues and to provide suggestions for dealing with those that directly affect the way we set up and conduct our research.

The book presupposes that we humans are reflexively conscious biological animals—that is, animals who are not only consciously aware, but aware of being aware. This means that our experience can be studied both as a biological phenomenon from the science side of the divide and as a subjective phenomenon from the humanistic side. The book is written for those interested in taking both perspectives into account to develop a naturalistic understanding of experiences deemed religious. Such a pursuit does not rule out religious understandings of experience that are compatible with a naturalistic approach, but it does not develop them. My own view is that the cultivation of some forms of experience that we might want to deem religious or spiritual can enhance our well-being and our ability to function in the world, individually and collectively. Identifying those forms, however, is not the purpose of this book.[2]

My eagerness to get on with the task is fueled by a long-standing set of interdisciplinary interests. Although trained as a historian of religion

---

[1] For a different approach to bridging the divide between the sciences and humanities that addresses humanistic fears more directly, see Slingerland (2008).

[2] For a discussion of spirituality from a naturalistic perspective, see Flanagan (2007) and Van Ness (1996).

with a particular focus on Christianity in the modern era, I was originally drawn into the field through discussions of theory and method, an interest I have maintained throughout my career. I was able to integrate those interests, or at least bring them into conversation with one another, in my book *Fits, Trances, and Visions*, which traced the history of the interaction between experiencing religion and medical and psychological explanations of experience over time.

Though this was not its overt focus, *Fits, Trances, and Visions* was inspired by the realization that there are commonalities between multiple personality, possession trance, and religious inspiration that are rooted in capacities of the mind, and that new insights could be generated by comparing the similarities and differences between them. This comparison, which has continued to fascinate me, led to further work across the disciplines of psychiatry, anthropology, psychology, and religious studies over the past decade and in the process generated the methodological reflections that make up the present book.

My particular interest in and preoccupation with unusual sorts of experiences has influenced the choice of examples presented in this book. There is no reason, however, why this bias should preclude using the approaches recommended here to study more ordinary types of experience. So, too, the traditions engaged reflect my own range of expertise. As the metaphor of rough and ready bridges is intended to suggest, I do not intend this book to be the last word on anything, including matters of method. I do hope, however, that it will foster a collaborative spirit among those interested in working across the humanities/sciences divide and an interest in testing and refining methods and theories in an effort to enhance our collective understanding of things deemed religious.

Religious Experience Reconsidered

The Problem of "Religious Experience"

The idea of "religious experience" is deeply embedded in the study of religion and religions as it (religion) and they (religions) have come to be understood in the modern West. In the nineteenth and twentieth centuries, many modernizers in the West and elsewhere advanced the idea that a certain kind of experience, whether characterized as religious, mystical, or spiritual, constituted the essence of "religion" and the common core of the world's "religions." This understanding of religion and the religions dominated the academic study of religion during the last century. Key twentieth century thinkers, such as Rudolf Otto, Gerardus van der Leeuw, Joachim Wach, Mircea Eliade, and Ninian Smart, located the essence of religion in a unique form of experience that they associated with distinctively religious concepts such as the sacred (Eliade 1957/1987), the numinous (Otto 1914/1958), or divine power (van der Leeuw 1933/1986).

This approach has been heavily criticized over the last thirty-five years on two major grounds. First, it sets religious experience up as the epitome of something unique or sui generis,[1] which must be studied using the special methods of the humanities. As a unique sort of experience, they argued that scholars should privilege the views of believers (the first person or subjective point of view) and should not try to explain their experiences in biological, psychological, or sociological terms for fear of "reducing" it to something else. Second, it constituted religion and the religions as a special aspect of human life and culture set apart from other aspects. Critics claimed that this approach isolated the study of religion from other disciplines (Cox 2006), masked a tacitly theological agenda of a liberal ecumenical sort, and embodied covert Western presuppositions about religion and religions (McCutcheon 1997; Sharf 1998; Fitzgerald 2000a; Masuzawa 2005).

The critics are basically right about this. Around 1900, that is, at the height of the modern era, Western intellectuals in a range of disciplines were preoccupied with the idea of experience (Jay 2005). This spilled over into theology and the emerging academic study of religion where thinkers with a liberal or modernist bent, mostly Protestant and a few

---

[1] "Sui generis" is a Latin phrase meaning "of its own kind." It refers to a person or thing that is unique, in a class by itself (*The New Dictionary of Cultural Literacy*, 3rd ed. 2002).

Catholic, turned to the concept of religious experience as a source of theological authority at a time when claims based on other sources of authority—ecclesiastical, doctrinal, and biblical—were increasingly subject to historical critique. For modernist theologians who followed in the steps of the liberal Protestant theologian Friedrich Schleiermacher, the self-authenticating experience of the individual seemed like a promising source of religious renewal, less vulnerable to the acids of historical critical methods (Proudfoot 1985; Sharf 1999; Jay 2005; Taves 2005).[2]

Early twentieth-century liberal Christian theologians, such as Rudolf Otto, Nathan Söderblom, and Friedrich Heiler, placed the experience of the numinous, sacred, or holy at the center of Christianity and, by extension, at the center of all other religions as well.[3] Hindu and Buddhist modernizers, such as Sarvepalli Radhakrishnan and Daisetz Teitaro Suzuki, made similar moves relative to their own traditions, using the idea of experience to undercut traditional sources of authority and interpret traditional concepts in new ways amidst the cross-currents of colonialism, westernization, and nationalist self-assertion. While maintaining the centrality of their own traditions, each used the notion of experience to underscore what they viewed as the essence of all religions (Taves 2005).

It was in this context that the Harvard psychologist William James gave the Gifford Lectures at the University of Edinburgh in 1902. These lectures, which were immediately published as *The Varieties of Religious Experience*, not only defined religion in terms of religious experience— that is, as *"the feelings, acts, and experiences of individual men in their solitude, so far as they apprehend themselves to stand in relation to whatever they may consider divine"* (James 1902/1985, 34)—but popularized what had been a predominantly Protestant concept as a core feature of religion in general (Taves 1999, 271). While James was responsible in many ways for initiating the turn to religious experience in the psychology of religion and religious studies, he did not—like so many who followed him—claim that religious experience was sui generis and refuse to explain it in psychological or sociological terms. Indeed, his aim as a

---

[2] We can and should distinguish between "religious experience" as an abstract concept, which has played a prominent role in modern religious thought, and "religious experiences" (in the plural) as specific behavioral events, which I refer to in what follows as "experiences deemed religious." The conflation of these two usages has created a great deal of confusion in the field.

[3] Otto, Heiler, and Söderblum were all Protestant theologians and early historians of religion, who followed the great liberal Protestant theologian Friedrich Schleiermacher in defining religion in terms of experience largely independent of doctrine and institution, resisted psychological interpretations of experience, and limited comparisons to religious phenomena.

psychologist was to explain religious experience in psychological terms, while at the same time leaving open the possibility that it pointed to something more (Taves, 2009a).

Although James should not be grouped with those who argued for a sui generis understanding of religion, his definition privileged experience of a particular sort over religious doctrine, practice, or institutions. In privileging sudden, discrete authenticating moments of individual experience (such as revelations, visions, and dramatic conversion experiences) over ordinary, everyday experience or the experience of groups, he introduced a bias toward sudden, individual experience that not only shaped the contemporary Western idea of religious experience but also related concepts such as mysticism and spirituality as well.

The prominent twentieth-century scholars of religion already mentioned—Gerardus van der Leeuw, Joachim Wach, Mircea Eliade, and Ninian Smart—built on this turn-of-the-century emphasis on experience to formulate their understanding of religion and the distinctive phenomenological methods they thought should be used to study it. In the wake of the general linguistic turn within the humanities, however, this entire approach was called into question. Many scholars of religion, eager to deconstruct an essentialist understanding of religion and religious experience, abandoned the focus on religious experience and recast the study of religion in light of critical theories that emphasize the role of language in constituting social reality in the context of relationships of power and inequality (Sharf 1998; Braun and McCutcheon 2000; Jensen 2003; Fitzgerald 2000b; McCutcheon 2002).[4] Scholars have now traced the history of these concepts in Western thought (deCerteau 1995; Jantzen 1995; Scharf 1999; Schmidt 2003, Jay 2005; Taves 2005), their appropriation by turn-of-the-century

---

[4] The linguistic or cultural turn refers to the application of insights drawn from linguistics, literary criticism, and cultural anthropology to a range of disciplines in the humanities, where it has been well received, and the social sciences, where it has been highly contested. This approach, which is part of a general postmodern critique, stresses the ways in which language shapes knowledge and treats all truth claims, including scientific ones, as forms of discourse that constitute social reality through relationships of power and inequality. In so doing, it privileges authorial virtuosity, while challenging natural and social scientific claims to generate generally valid, shared knowledge (Bonnell and Hunt 1999, 1–27). This shift in approach is evident in the entries on "Religion" in the first and second editions of the fifteen-volume *Encyclopedia of Religion*. In the first edition (1987), Winston L. King defined religion as "the organization of life around the depth dimensions of experience" and struggled to distinguish distinctively religious depth experiences from nonreligious ones, deciding rather circularly that "the religious experience is religious precisely because it occurs in a religious context" (2005, 7695–96). The supplementary entry written by Gregory D. Alles for the second edition (2005) highlights the intense criticism directed toward definitions of this sort and the scholarly shift from "trying to conceptualize religion to reflecting on the act of conceptualization itself" (2005, 7702) that ensued.

intellectuals with modernist inclinations in other parts of the world such as India and Japan (Halbfass 1988; Sharf 1995; King 1999), and their use by missionaries in colonial contexts (Chidester 1996, Fitzgerald 2007b). These studies, although well integrated with efforts at deconstruction across the humanities, are usually isolated from efforts to understand religion in the natural sciences. Indeed, those who embrace critical theory within the humanities and social sciences have typically been more interested in deconstructing scientific efforts than in bridging between science and critical theory (Wiebe 1999; Slingerland 2008).

Scholars in anthropology, sociology, and psychology—disciplines that we might expect to serve as bridges between the humanities and natural sciences—have faced various difficulties in that regard. Within mainstream anthropology of religion, the primary focus has been on shamanism and spirit-possession with far less attention paid to so-called world religions, particularly Christianity (Cannell 2006). In reciprocal fashion, religious studies has focused for the most part on "high religions" with "gods" and relegated the study of shamanism and spirits—that is, "folk religion"—to anthropology (Mageo and Howard 1996; Mayaram 2001). Although William James and his collaborators in the Society for Psychical Research thought of spirit-possession and mediumship as intimately related to the broader realms of religion and religious experience, they downplayed those connections in their published work and were not able to overcome the emerging division of labor between religious and theological studies, on the one hand, and the anthropology of religion, on the other (Kenny 1981; Taves, 2009a). Given this twentieth-century division of labor, scholars have tended to use terms such as "religious experience," "mysticism," and "spirituality" with reference to so-called "high" religions but not as commonly in relation to "folk" or "primitive" religion.

In terms of its orientation to the humanities and natural sciences, anthropology has been divided right down the middle. More than any other discipline, anthropology has been a battleground in the methodological wars between critical theorists oriented toward the humanities and social scientists oriented toward the natural sciences. While race and gender have been the most hotly contested issues, any attempt to bring science into the humanities and critical theory into science can raise suspicions among anthropologists (Slingerland 2008). There are pockets, however, within anthropology—psychological anthropology and medical anthropology in particular—that do bridge the humanistic and the natural sciences, and there is some exciting new work being done on religion in these subfields (e.g., Luhrmann 2004, 2005). Generally speaking, mainstream anthropological research on shamanism and spirit-possession has exemplified the tension between reductionistic, naturalistic, or medical models, on the one hand, and phenomenological, contextualizing cultural-studies approaches, on the other (Boddy 1994), although here, too, a few anthropologists have made innovative efforts to bridge the gap between the natural sciences and the humanities (Stephen 1989).

Although some sociologists, especially those following in the tradition of Emile Durkheim, have attended to collective and in some cases even individual experience, they have focused on the social causes and effects of experience apart from the psychological and biological. In general, psychologists and sociologists of religion have distinguished between the private religious experience of individuals and the public religiosity of organized groups, with psychologists of religion focusing on the former and sociologists of religion on the latter. Although there is some newer work (e.g., Bender 2008) that runs counter to these trends, psychologists of religion have devoted far more attention to religious experience than sociologists.

Due to their focus on religious experience, including spirituality and mysticism, the relationship between the psychology of religion and the general field of psychology is parallel in some respects to the relationship between religious studies and other disciplines. Although research in the psychology of religion is conducted across the whole array of subfields within psychology running the gamut from the natural to the social sciences (see Paloutzian and Park 2005), psychologists of religion, like scholars of religion, have wrestled with the question of whether religion is unique among human behaviors or can be accounted for using the research methods and/or explanatory principles that are applied to human behavior more generally (Baumeister 2002). Those who claimed that religion is in some sense unique (sui generis) have resisted "reductionistic" approaches to the psychology of religion and maintained the need for distinct approaches that set it apart from the rest of psychology (Dittes 1969, Pargament 2002).

While the psychology of religion, like religious studies, has been through a long period of critical self-reflection, some within the field now advocate a "multilevel interdisciplinary paradigm" (Emmons and Paloutzian 2003) that would allow the psychology of religion to "reach out to evolutionary biology, neuroscience, anthropology, cognitive science, and . . . philosophy in a generalized cross-disciplinary approach to critiquing and sharpening the assumptions of science" (Paloutzian and Park 2005a, 7–9). The multilevel interdisciplinary paradigm would thus link "subfields within psychology as the core discipline in a broader effort." This new paradigm undercuts the old binary distinction between reductionism and uniqueness, reframing it in relation to theories of emergence in which emergent properties, such as consciousness and group leadership, are understood to emerge at different levels of analysis (ibid). Experience—whether religious, spiritual, or mystical—is definitely a phenomenon for study within this new paradigm, but the implications of the paradigm for setting up experientially related objects of study that can be examined across disciplines have not been adequately worked out. Without further refinement at the design stage, it will be difficult to connect different lines of research.

Finally, in the last decade and a half (since 1990), there has been a dramatic increase in studies examining the neurological, cognitive, and evolutionary underpinnings of religion in light of the rapid advances in the study of the brain and consciousness. Scholars who identify with the growing subfield of the cognitive science of religion are drawn from disparate disciplines including psychology, anthropology, religious studies, and philosophy. Though most of them are well versed in the study of religion, they have focused on belief and practice (ritual) and with a few exceptions, such as Azari (2004) and Livingston (2005), have ignored experience (for an overview, see Slone 2006). In addition, scholars and researchers, including a number of self-identified neurotheologians, most of whom lack training in theology or religious studies (e.g., D'Aquili and Newberg 1999), have enthusiastically embraced the challenges of identifying the neural correlates of religious experience without engaging the critiques of the concept that led many scholars of religion to abandon it.

After decades of critical discussion of the concept, we can neither simply invoke the idea of "religious experience" as if it were a self-evidently unique sort of experience nor leave experience out of any sensible account of religion. How, then, should we understand "religious experience"? Given the critiques of the last several decades, is there any way the concept can be studied by those interested in understanding such experiences naturalistically?

EXPERIENCES DEEMED RELIGIOUS

Rather than abandon the study of experience, we should disaggregate the concept of "religious experience" and study the wide range of experiences to which religious significance has been attributed. If we want to understand how anything at all, including experience, *becomes* religious, we need to turn our attention to the processes whereby people sometimes ascribe the special characteristics to things that we (as scholars) associate with terms such as "religious," "magical," "mystical," "spiritual," et cetera. Disaggregating "religious experience" in this way will allow us to focus on the interaction between psychobiological, social, and cultural-linguistic processes in relation to carefully specified types of experiences sometimes considered religious and to build methodological bridges across the divide between the humanities and the sciences.

A focus on things deemed religious in turn allows us to make a distinction between *simple ascriptions,* in which an individual thing is set apart as special, and *composite ascriptions,* in which simple ascriptions are incorporated into more complex formations, such as those that scholars and others designate as "spiritualities" or "religions." This distinction provides a basis for examining the various roles that experience in general and unusual experiences in particular play in both *simple ascriptive formations* (in which, e.g., a single event is set apart as special) and *composite ascriptive formations* (in which, e.g., an event is viewed as originary and people seek to recreate it in the present).[5] The distinction between simple and composite formations, thus, allows us to envision a way of studying "religion" that allows us to understand how humans have used things deemed religious (simple ascriptions) as building blocks to create the more complex formations (composite ascriptions) we typically refer to as "religions" or "spiritualities."

Previous Work

The distinction between simple and composite ascriptions builds on a particular reading of the French sociologist Emile Durkheim and has been anticipated to varying degrees in more recent work. James Dewey (1934) anticipated a similar distinction when he stressed the difference between "religion, a religion, and the religious" (3) and referred to "religious elements of experience" rather than "religious experience" in order to avoid setting up religious experience as "something sui generis" (10, 13). More recently, Hent de Vries (2008, 11–12) makes an analytical distinction between a general or generic concept of religion and the "things" (words, gestures, powers, et cetera) that constitute the "elementary forms" in which religion, abstractly conceived, is instantiated. De Vries's approach in *Religion: Beyond a Concept,* like Dewey's in *A Common Faith,* is constructive as well as analytical. Where Dewey sought to articulate a scientifically grounded "common faith," de Vries and collaborators seek to move beyond the abstract concept of religion to develop what he calls a "negative metaphysics" or "minimal theology" designed to sketch the "emerging features" of an "abstract and virtual 'global religion'" (de Vries 2008, 13).

---

[5] I refer to simple and composite formations rather than simple and composite ascriptions when I want to encompass the beliefs and practices that are associated with a simple or composite ascription. References to simple and composite formations should always be understood to mean simple and composite *ascriptive* formations. These and other italicized terms can be found in the glossary.

The strictly analytical focus of the distinction made here, however, more closely parallels that of sociologist Danièle Hervieu-Léger and psychologists such as Kenneth Pargament and Annette Mahoney. Hervieu-Léger distinguishes between the sacred character that can be conferred on things and religion as a way of organizing meaning through chains of belief (Hervieu-Léger 2000, 106–8). Pargament and Mahoney (2005, 180–81) distinguish between the sanctification of various objects or aspects of life and religion as "a search for significance in ways related to the sacred." In making these distinctions, these scholars redefine the *first-order terms* "sacred" and "religion" as *second-order terms* for the purposes of their research.

Although I adopted this course as well in earlier drafts, doing so makes it harder to distinguish between our aims as scholars and those of the people we are studying and thus risks obscuring the contestations over and transformations of experience that we want to study on the ground. Since there is no way to specify an inherently contested phenomenon precisely, I will propose that we situate what people variously refer to *emically* (on the ground) as "religious," "spiritual," "mystical," "magical," and so forth in the context of larger processes of meaning making and valuation, and specifically in relation to the process of *singularization* (Kopytoff 1986), by means of which people deem some things special and set them apart from others. In my revisions, I have tried to be clear rather than relentlessly consistent in my use of terms, so the reader will find references to both "experiences deemed religious" and "things considered special" as seems appropriate in any given context.

The distinction between simple and composite ascriptions relies heavily on *attribution theory*, which seeks to explain how people explain events. Long a staple of social psychology, attribution theory was applied to religious experience in the 1970s (Proudfoot and Shaver 1975; Proudfoot 1985) and to religion in general in the 1980s (Spilka, Shaver, and Kirkpatrick 1985). Although attribution theory has been widely presupposed by psychologists of religion (Spilka and McIntosh 1995), some religious-studies scholars have rejected it because they think it overrides the subjective sense of those who claim their experiences are inherently religious rather than culturally constructed (Barnard 1992; Barnard 1997, 97–110). In order to respond to this criticism, we will need to distinguish between *attributions* (commonsense causal explanations) that people often supply consciously and *ascriptions* (the assignment of a quality or characteristic to something) that may be supplied implicitly below the threshold of awareness.[6] The distinction between attribution and ascrip-

---

[6] In everyday speech, the terms "ascription" and "attribution" are typically used interchangeably to refer to both causal explanations and the assignment of a quality or charac-

tion will allow us in turn to connect attribution theory more fully with research on implicit, nonconscious mental processing.

Nina Azari, one of the few neuroscientists with dual doctorates in both psychology and religious studies, appropriates attribution theory critically in her recent religious-studies dissertation (Azari 2004). The dissertation, which builds on her pioneering use of brain-imaging techniques to identify neural correlates of religious experience (Azari et al. 2001; Azari, Missimer, and Seitz 2005), provides the most sophisticated attempt so far to come to terms with the issues surrounding the neuroscientific study of religious experience. While our conclusions are compatible, they are intended for different audiences and thus are expressed in somewhat different terms and framed at different levels of generality. Azari's work is directed primarily toward neuroscientists studying contemporary Western subjects, philosophers of religion, and theologians interested in reflecting on their findings. She critiques both attribution theory and the relatively unsophisticated theoretical underpinnings of neuroscientific studies of religious experience in light of recent research on emotion. This research allows her to undercut the inadequate conceptualization of emotion that informed earlier neuroscientific studies of religious experience as well as overly narrow conceptions of causality in some versions of attribution theory (Azari 2004, 172–82). In contrast, this book aims to rehabilitate a more broadly defined concept of experience and to suggest an approach to studying experiences deemed religious that can be used by researchers who do not focus on contemporary Western subjects.

Azari's approach has specific limitations that need to be overcome in order to advance this larger agenda. First, although defining religious experience from the perspective of the subject works well when studying modern Western subjects for whom the concept of religious experience is meaningful, this work aims to support research on singular experiences across cultures and historical time periods. Second, although experience can usually be construed as having an emotional valence, it is not always its most salient feature. Defining experience in terms of emotion deflects attention from a range of unusual experiences that are granted special significance, such as lucid dreams, auditory and visual hallucinations, sensed presences, possession trance, and out-of-body experiences, which this book seeks to include. Third, a focus on individual, decontextualized

---

teristic to something. In the context of attribution theory, however, social psychologists use "attribution" to refer specifically to the commonsense causal explanations that people offer for why things happen as they do (Försterling 2001, 3–4). In what follows, I distinguish between attribution and ascription, using "attribution" to refer to causal explanations, as in social psychology, and using "ascription" to refer to the assignment of a quality or characteristic to something.

experiences tends to reproduce the relatively narrow understanding of "religious experience" that has been of particular interest to modern Western philosophers of religion and theologians. By extending attribution theory to processes at the group level and to composite as well as simple ascriptions, we can place the study of experiences that people consider special within a broader interdisciplinary field of inquiry and open new possibilities for understanding the way that religions are constructed.

THE ARGUMENT

The argument unfolds in chapters devoted to religion, experience, explanation, and comparison. Chapter 1 (*Religion*) addresses the question of how scholars can specify what it is they want to study without obscuring the contestations over meaning taking place on the ground. Since there is no way to specify an inherently contested phenomenon precisely, I argue that scholars can situate what people characterize as religious, spiritual, mystical, magical, superstitious, and so forth in relation to larger processes of meaning making and valuation, in which people deem some things special and set them apart from others. We can then identify marks of specialness (that set things apart in various ways), things that are often considered special (ideal things and anomalous things, including anomalous beings), and the ways in which simple ascriptions of specialness can be taken up into more complex formations. These various distinctions provide numerous options for setting up more precisely designed research projects to probe competing schemes of valuation and singularization in different social contexts.

Chapter 2 (*Experience*) reconsiders views of experience and representation that have colored humanistic discussions of religious experience in light of recent discussions of experience and consciousness among philosophers, psychologists, and neuroscientists. Distinctions between types of consciousness (transitive and intransitive), levels of consciousness (lower and higher), and levels of mental processing (conscious and unconscious) allow us to consider the relationship between experience and representation in an evolutionary and developmental perspective relative to the experience of animals and prelinguistic humans. Viewing experience in this way allows us to consider how we gain access to experience (our own and that of others) and how it acquires meaning as it arises in the body and through interaction with others. A more dynamic model of how we articulate our own experience and that of others illuminates a range of data that we can gather about experience and allows us to reconsider the relationship between experience and representation in some specific cases (dreams, possession trance, and meditation) in light of the data available for studying them.

The dynamic model of how we come to know our experience and the experiences of others developed in chapter 2 is based on research on *embodiment* and *theory of mind*. Theory of mind is a key aspect of what researchers refer to as "folk psychology," the set of very basic, cross-culturally stable assumptions that we use to predict, explain, or understand the everyday actions of others in terms of the mental states we presume lie behind them. Folk psychology, which also informs the latest work in attribution theory (Malle 2004, 2005), lays the foundation for the development of a more interactive understanding of how and why people explain their own and other's actions in chapter 3 (*Explanation*). Drawing on the multilevel attributional framework proposed by Hewstone (1989), I show how Malle's interactive approach can be extended to various levels of analysis—intrapersonal, interpersonal, intragroup, and intergroup—some of which fall under the traditional purview of historians and ethnographers. Though the attributional process takes a somewhat different form at each level, an interactive approach allows us to conceptualize everyday explanations as an interpretive process involving negotiation and contestation at every level.

In arguing against the sui generis approach to religious experience, I am arguing that the comparison between religious and nonreligious subjects taken for granted in experimental design can and should be extended to historical and ethnographic research. In chapter 4 (*Comparison*), I sketch some of the ways that researchers can construct similar sorts of comparisons using historical and ethnographic data. Returning to the distinctions between simple and composite formations set out in chapter 1, I set up comparisons that illustrate what we can learn from comparisons between simple formations, between composite formations, and between simple and composite formations.

The distinctions between ascription and attribution and simple and composite formations have implications not only for the study of experiences that people consider special but also for the study of religion more generally. The distinction between ascription and attribution allows us to distinguish between the creation of special things through a process of singularization, in which people consciously or unconsciously ascribe special characteristics to things, and the attribution of causality to the thing or to behaviors associated with it. The distinction between simple ascriptions, in which an individual thing is set apart as special, and composite ascriptions, in which simple ascriptions are incorporated into the more complex formations characteristic of religions or spiritualities, in turn allows us to envision a building-block approach to the study of religion. The implications of these distinctions for the study of religion are drawn out in the conclusion.

WHY AN ATTRIBUTIONAL APPROACH IS BETTER

Reframing the concept of "religious experience" initially as "experiences deemed religious" and then more broadly as a subset of things people consider special allows us to do three things. First, it forces us to sort out who is deeming things religious or characterizing them as special and on what grounds, both at the level of scholarship and that of general human behavior. Analysis of the different ways that things can be set apart as special and protected by taboos will suggest that the sui generis approach to the study of religion, which defines religion in terms of religious experience, sets the study of religion apart and protects it with taboos against comparing it with nonreligious things. If instead we situate the processes whereby people characterize things as religious, mystical, magical, and so forth within larger processes of meaning making and valuation (singularization), we are better able to analyze the contestations over the meaning and value of particular things and the way that those things are incorporated into and perpetuated by larger socio-cultural formations, such as religious traditions and spiritual disciplines.

Second, it allows us to position experience, traditionally understood as a central concept within the study of religion, not as something that sets the study of religion apart from all other forms of knowledge but rather locates it in relation to them. By locating how we come to know our own and others' experience through processes that are simultaneously embodied and interactive, we can make a concept familiar to scholars of religion usable across disciplines and further a process of conceptual integration that is presupposed in the natural sciences but less well advanced elsewhere.[7] In drawing from different disciplines to examine processes of ascription and attribution at and between various levels (intrapersonal, interpersonal, intragroup, and intergroup), we can escape the simple binaries in which the reductionism debate has been framed in religious studies and explore the distinctive features of different levels of analysis in more sophisticated ways.

[7] While vertical integration across levels of analysis is taken for granted in the natural sciences, this is not the case in the social sciences or the humanities (Slingerland 2008). Calls for integration across levels and more sophisticated analysis of interactions between levels are becoming more common in the social sciences and the humanities (see, for example, Barkow, Cosmides and Tooby 1992 [in evolutionary psychology]; Emmons and Paloutzian 2003; Paloutzian and Park 2005, Kirkpatrick 2005 [in psychology of religion]; and Clayton 2004; Clayton and Davies 2006 [in religious studies]).

Third, an attributional approach allows us to view experiences—and especially unusual experiences—as a subset of the many special things that may be incorporated into the more complex formations we think of as "religions." The twentieth-century focus on "religious experience" rather than experiences deemed religious deflected attention from the various components that taken together constitute a "religion." Refocusing our attention on the component parts and the disparate ways in which they can be assembled provides a method for assessing the role of unusual experiences in the emergence and development of religions. Although conceived to solve the problems surrounding "religious experience," the method provides a more promising way forward for the study of religion generally.

CHAPTER ONE

Religion

DEEMING THINGS RELIGIOUS

Key figures associated with the emergence of the scholarly study of religion disagreed sharply over how sacred or holy or religious things ought to be characterized and, by extension, how they could be understood. Rudolf Otto, a German theologian and historian of religions, argued that the holy should be characterized in terms of a distinctive nonrational element, which he called "the numinous." This distinctive numinous object gave rise to an associated feeling or mental state that Otto claimed was "perfectly sui generis and irreducible to any other." As such, it could not be precisely defined and certainly could not be explained in terms of other, more ordinary feelings. The only way to help others understand it, he said, was to discuss how it was like and unlike other things until they began to experience it for themselves (Otto 1923/1958, 7).

In *The Varieties of Religious Experience*, the American psychologist William James made the opposite claim. He argued, by way of contrast, that as far as he could tell there were no distinct religious emotions, such as Otto's feeling of the numinous. Moreover, he speculated that there might not be any specifically religious objects or essentially religious acts. He viewed religious emotions as composites that could be broken down into an ordinary feeling and an associated religious concept (James 1902/1985, 33). The French sociologist Emile Durkheim elaborated this idea more fully in *The Elementary Forms of the Religious Life*, wherein he defined a religion as "a unified system of beliefs and practices relative to sacred things, that is, things set apart and forbidden" (Durkheim 1912/1995, 44). Here, as William Paden observes, the sacred is that which is set apart and forbidden. As such it is purely relational and has no essential content of its own. "The sacred is simply what is *deemed* sacred by any group" (Paden 1994, 202–3 [emphasis in original]).

Before proceeding, it is important to note that in the space of two paragraphs we have had occasion to refer to numinous objects (Otto); religious experiences, objects, and acts (James); and sacred things (Durkheim). Moreover, Durkheim distinguished between "sacred things" and "a religion," which he understood as "a unified system of beliefs and practices relative to sacred things." In what follows, I will use "things" to refer to *any* thing, whether an experience, object, act, or agent.

This chapter treats "religious experience" as a kind of "religious thing." I will consider the merits of two basic approaches to "religious experience," one in which the religiousness of the experience is understood to be inherent (sui generis) in the experience itself and the other in which it is viewed as ascribed to it. I will opt for the latter approach while acknowledging two major difficulties that must be overcome: first, the difficulty that scholars face in specifying what we mean by "religious" and, second, the difficulty posed by subjects' claims that some experiences seem inherently religious, spiritual, or sacred. Deferring this second difficulty to later chapters, I argue that the use of "religious" or any other first-order term, such as "numinous," "sacred," "mystical," "spiritual," or "magical" as a means of specifying an object of study is both limiting and confusing and suggest instead that investigation of the broader, more generic category of "special things" and "things set apart" may be more helpful for the purposes of research. Building on Durkheim, I distinguish between things that people view as special or that they set apart, on the one hand, and the systems of beliefs and practices that some people associate with some special things, on the other. The former involves a simple ascription (of specialness) and the latter a composite ascription (of efficacy to practices associated with special things) characteristic of what we think of as *religions* or *spiritualities*.

The Sui Generis and Ascriptive Models of "Religious Experience"

The two approaches to the study of "religious experience," which I will refer to as the sui generis model and the ascription model, are summarized in table 1.1. They differ over whether there are uniquely religious (or mystical or spiritual) experiences, emotions, acts, or objects. The sui generis model assumes implicitly or explicitly that there are. The ascriptive model claims on the contrary that religious or mystical or spiritual or sacred "things" are created when religious significance is assigned to them. In the ascriptive model, subjects have experiences that they or others deem religious. One of the ways that ambiguity is maintained with respect to the two models is by referring to "religious experience," as if it were a distinctive thing, rather than using the more awkward, but clearly ascriptive, formulation, "experiences deemed religious."

There are significant methodological implications of the seemingly minor shift from "religious experience" to "experiences deemed religious" for both the explanation and comparison of religious phenomena. Most of the scholarly discussion has focused on whether religious phenomena can or should be explained in nonreligious terms, with those who advocate a

TABLE 1.1

Methodological Differences between the Sui Generis and Ascription Models

Guiding Question	Model	
	Sui generis	Ascription
Are some experiences inherently religious?	Assumes that there are some things (most often experiences) that can be viewed as inherently religious or mystical.	Assumes that things (events, experiences, feelings, objects, or goals) are not inherently religious or not-religious but must be constituted as such by persons.
Should certain things always be considered religious?	Yes. There are underlying things (again often experiences), which can or should be understood as (authentically) religious or mystical or spiritual.	No. Diverse things can be deemed religious—"mysticism" is a modern category—and there are diverse views regarding what should "count" as religious, mystical, or spiritual.
What can be compared?	Religious things are compared with other religious things. Common features are often granted evidential force relative to religious claims.	Experiences are compared with other things that have some similar feature(s) whether they are viewed as religious or not.
What is the goal of comparison?	To understand more about religious or mystical things.	To understand how and why people deem things religious and allow researchers to explore the making and unmaking of religion.
How do they relate to other things?	Religious experience is set apart from other experiences and tacitly protected from comparison with them.	Experiences deemed religious are viewed in relation to other experiences and subject to comparison with them.

sui generis approach arguing that they should not. Indeed, for most scholars, the claim that religion is sui generis is simply another way of saying that religion cannot or should not be explained in anything other than religious terms. During the latter part of the twentieth-century, the idea that religion is sui generis was advanced as a "disciplinary axiom" (Pals 1986, 35; 1987, 269). Stated positively, it asserted that religious things must be explained in religious terms; negatively, it prohibited "reducing" religion to something else by explaining it in *nonreligious* terms.[1]

In their focus on the issue of reductionism, scholars in the sui generis camp largely overlooked the way in which their model affected how scholars of religion set up and utilized comparisons. Even though individual scholars within this camp—such as Otto—advocated comparison, the logic of the sui generis position has more commonly led to assertions of the incomparable nature of uniquely religious things (Smith 1990, 36–53). Religion scholars under the sway of this position tend to limit themselves to comparing different religious things. By contrast, the ascriptive model frees us to compare things that have features in common, whether they are deemed religious or not. Doing so allows us to focus on how and why people deem some things, including some experiences, as religious and others as not.

The ascription model sketched here is inspired by research within the field of social psychology on attribution theory, which explains how people explain events. In everyday speech, the terms "ascription" and "attribution" are typically used interchangeably to refer to both causal explanations and the assignment of a quality or characteristic to something. In the context of attribution theory, however, social psychologists use "attribution" to refer to causal attributions: that is, to the commonsense explanations that people offer for why things happen as they do (Försterling 2001, 3–4). If we follow their lead and limit the use of "attribution" to causal explanations, we can use "ascription" to refer to the assignment of a quality or characteristic to something and use this distinction to further clarify the differences between the sui generis and deeming models.

---

[1] A number of scholars have worked to clarify more precisely what the term has meant in practice for religious studies (Segal 1983; Wiebe 1984; Pals 1986, 1987; McCutcheon 1997; Pyysiäinen 2004). Pyysiäinen, after reviewing a number of possible meanings, concluded that in practice "regarding religion as sui generis is merely an apriori strategy for not decoupling religious beliefs from their religious metarepresentational context and for treasuring religious experience as something irreducible" (2004, 80). Pals made something of the same point some years earlier when he observed that major midcentury figures in the study of religion (for example, Gerardus van der Leeuw, William Brede Kristensen, Joachim Wach, Mircea Eliade, and C. J. Bleeker), unlike some of their predecessors, made no attempt to argue for a sui generis understanding of religion on psychological or philosophical grounds, but "simply laid [it] down, apodictically as it were, . . . as the 'principle' which guides scholarly inquiry" (1987, 269). Much of the discussion of the limitations of a sui generis approach has centered on the work of Mircea Eliade, who was perhaps the most influential figure in the establishment of religious studies as a field in American universities in the 1960s and 1970s. Although Eliade has been widely criticized for promoting a view of the sacred as sui generis and ontologically autonomous, some who know his work best argue for a more nuanced view (see Rennie 1996, 179–212).

The central question that divides the two models is whether religious things, existing as such, have special inherent properties that can cause things to happen or, alternatively, whether people characterize things as religious and thus endow them with the (real or perceived) special properties that are then presumed to be able to effect things. In the sui generis model, it is assumed that religious things exist and have inherently special properties. In the ascription model, it is assumed on the contrary that people ascribe religious characteristics to things to which they then attribute religious causality. The ascriptive—or deeming—model, thus, makes a fundamental claim about ascription (the assignment of qualities or characteristics) prior to attributing causality.

Arguments against the Ascription Model

Scholars have attempted to challenge the ascriptive model by invoking experiences that subjects feel are inherently religious and that have recurred in similar forms across time and cultures The two most common claims made in this regard are that the common features of unusual experiences point to the transcendent and that certain experiences are inherently mystical or religious.

Common features point to the transcendent. Many philosophers of religion with an interest in religious experience recognize a variety of different types of religious experience, but two types—mystical and numinous—are frequently singled out for attention. Although there are also various definitions of these two terms, "mystical" is often used to refer to experiences of unity with or without a sense of multiplicity and "numinous" to experiences of a felt presence whether loving or fearsome. These differences notwithstanding, some philosophers argue that such experiences constitute a "common core" of religious experience and point out, not coincidentally, that these experiences are the ones that are most difficult to explain in naturalistic terms (Davis 1989, 176–77, 190–91, 233). Hood (2006) claims that psychological measurement based studies provide evidence for a common core of mystical experience, though he does not use these data to argue for or against the transcendent.

Certain experiences are inherently mystical or religious. So-called numinous and mystical experiences are the two types of experiences that people are most likely to consider inherently religious (Hood 2005, 356–60). The most serious criticism directed specifically toward traditional attribution theory is that it tends to override the views of religious subjects who understand their experiences in this way (Barnard 1992; 1997, 97–110). Attribution theorists must be able to account for experiences that individuals feel are inherently religious or mystical and that seem to bear little relationship to the subjects' preexisting beliefs or context (Davis 1989, 232–35). To account for this sort of subjective feeling, an ascriptive model must be able to show how implicit religious ascriptions can be built into experiences through preconscious mental processes in such a way that subjects feel they recognize or discover—rather than ascribe characteristics to—the experiences they consider religious or mystical.

Problems with the Sui Generis Model

The basic problem with the sui generis model is that it obscures something that scholars of religion should be studying: that is, the process whereby people constitute things as religious or not. Earlier practitioners of the sui generis model, the classical phenomenologists of religion, obscured this process by essentializing the bond between the "thing" and the religious ascription. Today, many scholars of religion do so by limiting their research to phenomena they deem religious, rather than investigating when people directly involved with the "thing" in question deem it religious or not.

Experimental researchers perpetuate this problem when they use a common-core model of religious experience to interpret their data. In a series of studies in which they measured changes in brain activity during meditation using single photon emission computed tomography (SPECT), Newberg (Newberg et al. 2001a) found that their scan images showed unusual activity in the area of the brain—the posterior superior parietal lobe—that may be responsible for orienting the individual in physical space and, thus, for maintaining the distinction between self/not self. In a popular presentation of their research, Newberg (Newberg et al. 2001b) writes, "As our study continued, and the data flowed in, Gene [D'Aquili] and I suspected that we'd uncovered solid evidence that the mystical experiences of our subjects—the altered states of mind they described as the absorption of the self into something larger—were associated . . . with a series of observable neurological events." This led them to wonder if they had "found the common biological root of all religious experiences" (2001b, 4–5, 9).

Whether or not they have found such a biological root depends entirely on how they define "religious experience." Newberg and D'Aquili have chosen a definition that emphasizes the dissolution of the sense of self and other. If their experimental results hold up, and if they define religious or mystical experience in this way, the answer could well be yes. Had they posed their research question in the ascriptive mode—had they asked, "have we found the common biological root of all experiences deemed religious?"—the situation would appear quite different. For anyone with even a rudimentary awareness of the wide range of experiences that humans have deemed religious or even the range that philosophers of religion, such as Davis (1989), have deemed mystical and numinous, the answer to this latter question would have to be no. Had they been working ascriptively, they could not simply have applied a definition, but would have had to distinguish more carefully between the way their subjects described their experience—including whether they described the experiences as religious or mystical—and the way they, as researchers, described the experiences.

Their popularized discussion is problematic not only because they use a common-core model of religious experience but also because their generalizations are based only on data from religious subjects (Tibetan Buddhist and Roman Catholic meditators). All they can report (and all they do report in their scientific articles) is an association between the altered states of mind that their subjects described as the absorption of the self into something larger and unusual activity in the posterior superior parietal lobe of the brain. Future studies should compare meditators in various religious contexts with those who have had similar experiences in *nonreligious* contexts. Paying careful attention to how experiences are described and who, if anyone, describes them as religious, would allow us to compare experiences in which the self feels it is absorbed into something larger in terms of (1) underlying neurological events, and (2) the conditions under which religious significance is attached to such events.

The ascriptive model thus allows us to say both less and more than the sui generis model. It allows us to say less in the sense that it forces us to eschew claims about religious experience, mystical experience, or spiritual experience in general and to identify specific aspects of experience that are sometimes deemed religious and *sometimes not*. What we gain in return is a way to get at the range of experiences sometimes deemed religious and to understand some of the variety of neural events and psychological processes that inform them.

DEEMING THINGS RELIGIOUS

To make the switch from a sui generis to an attributive formulation useful, we need to specify the type of experience to be studied, who does the "deeming," and what is meant by "religious." Before addressing the first question—that of experience—we need to address the other two, which are interrelated. Even if our primary interest is in how people on the ground deem things religious—that is, in what counts as religious for them—we still need to specify what we mean by "religious," if we are not going to limit our study to people who use that specific term or some easily recognizable cognate. If we want to compare ascriptions across cultures and time periods, how do we specify the kind of ascriptions that interest us? This question opens up problems that scholars of religion have discussed at great length without reaching any clear resolution. To get at this, we need to start by clarifying the difficulties that scholars have identified.

Clarifying the Difficulties

Put simply, scholars interested in understanding the processes whereby people in everyday life deem things religious are in a bind. "Religion," as scholars regularly point out, does not designate a specific, cross-culturally stable thing that we can reliably look for on the ground. Any specification of "religious" (or "spiritual" or "mystical" or "sacred" or "magical"), whether by scholars or practitioners of religions or believers who are the subject of scholarly investigation, excludes phenomena that some people sometimes deem religious and includes other things that most would not consider religious. Moreover, as an abstract Western concept, religion has been defined and refined in contexts of intercultural engagement in which the power to categorize and define has typically been unequal (Braun 2000; Fitzgerald 2000a, 2007b). Some anthropologists have even suggested that religion is basically a folk category indigenous to Western culture that has come to represent "a great potpourri of ideas and behavior with many independent evolutionary origins outside religion itself" (Dow 2007; Saler 2000, ix–x, 21–23). Scholars have acknowledged these difficulties and by and large have conceded the impossibility of specifying a uniquely religious set of phenomena viable across cultures and time periods.

Given the impossibility of specifying a scholarly definition valid for all times and places, some have suggested that we focus instead on how our subjects define religion (Asad 1993; Arnal 2000, 30–33). Others have proposed that we abandon the term in favor of concepts more immediately relevant to our research context (Fitzgerald 2000a). While agreeing that essentialist definitions must be rejected, most scholars in religious studies and the psychology of religion favor a third option: precise stipulation of what they as scholars mean by religion in the context of their research (Smith 1998; Braun 2000, 6–10; Zinnbauer and Pargament 2005; Tweed 2006, 29–53).

Each of these three options has strengths and liabilities. The first two strategies retreat from general formulations to more localized or indigenous usage. The first strategy attends to subjects who use the term "religion" or easily recognizable cognates: that is, subjects most commonly, though not exclusively, found in modern Western contexts. The second strategy, elaborated in contexts where "religion" or related cognates are not used, defines an object of study in terms that are more closely related to local usage in non-Western and/or pre-modern contexts. Neither strategy, however, facilitates comparative work across time and cultures.

When scholars specify what they mean by "religion" for purposes of research, they are trying to get around this problem. While I, too, want to identify concepts that facilitate comparisons across time and culture, I agree with the critics who argue that "religion" presents numerous difficulties for identifying ascriptions. The chief difficulties are as follows:

1. "Religion" is an abstraction that elides the distinction between the adjectival (things deemed religious, sacred, mystical, et cetera) and nominative (religions, spiritualities, paths, et cetera) use of terms and thus obscures the relationship between simpler and more complex phenomena. Though a few scholars have followed Durkheim in stressing this or similar distinctions (most notably Hervieu-Léger 2000, 101–20), a widespread tendency to overlook this distinction has hampered analysis.[2] An ascriptive approach highlights these differences. Viewed ascriptively, we can see that the study of religion can focus on either of two steps in a two-step process, in which, first, some things are deemed religious and then, second, some things deemed religious are made into religions. To put it another way, this means:

- *Religion* is an abstraction. It is not the same thing as *a religion* (singular) or *religions* (plural).
- A *thing that is deemed religious* (or sacred or magical) is not the same as *a religion*.
- We can ask when, how, and why things are deemed religious. We can ask if, how, and when a thing that is deemed religious is incorporated into or becomes a religion, a spirituality, or a path.

2. Scholars often equate defining "religion" with defining their object of study, thus limiting their object of study to what they or their subjects have defined as "religious." Differentiating between the "thing" we are studying and the deeming of it as "religious" highlights two features

---

[2] Fitzgerald provides an interesting example of this confusion. Lamenting the scholarly tendency to conflate "religion" and the "sacred," Fitzgerald (2007a, 72) cites what he refers to as Durkheim's definition of "religion" as illustrative of this conflation: "All known religious beliefs . . . present a common quality: they presuppose a classification of things—real or ideal things that men represent for themselves—into two classes, two opposite kinds, generally designated by two distinct terms effectively translated by the words *profane* and *sacred*. The division of the world into two comprehensive domains, one sacred, the other profane, is the hallmark of religious thought. (Durkheim 2001, 36)." This is not in fact a definition of "religion," but rather a definition of "religious beliefs," one of "the elementary phenomena that generate any religion" (Durkheim 1912/2001, 36), or, in the Fields translation, one of "the elementary phenomena from which any religion results" (Durkheim 1912/1995, 33).

that we need to specify: the point of analogy that defines the set of the things we are going to study and what we mean by a religious ascription. Thus, for example, we may choose to focus on experiences (that is, "things," broadly defined) that share a stipulated point of analogy such as a feeling of peace, engagement with an unseen entity, or loss of self-world boundaries. If we do not assume that these experiences are inherently religious, then our object of study becomes *things that share a stipulated point of analogy* as they intersect with meaning-making processes that lead to their characterization as religious or not.

3. Scholars often conflate their definition of religion with that of their subjects, failing to fully appreciate the diversity of and contestations over what counts as religion on the ground. This tendency is exacerbated by the use of "religion," "religious," and "religions" as both first- and second-order terms. As first-order terms, they are Western folk concepts, albeit ones with increasingly global usage. Their meanings, which are unstable and contested, overlap in some cases and compete in others with a variety of other terms derived from both Greek and Latin, including "magical," "religious," "sacred," "superstitious," "worshipful," "possessed," "insane," "inspired," and "secular." No matter how carefully scholars define what they mean by "religion," "religious," or "religions," it is hard to maintain a clear distinction between first- and second-order use of the terms when both appear in the same scholarly work. While there may be strategic value in using "religious" as a second-order term, especially for scholars of religion located in departments of religious studies, its use as both a first- and second-order term risks conflation and confusion of levels.[3]

Given that "religion," "religious," and "religions" are Western folk concepts, that their meaning is unstable and contested, and that they cannot be defined so as to specify anything uniquely, we need to consider

[3] Use of the term is intimately bound up with questions of disciplinarity, specifically whether and on what grounds "religious studies" should be constituted as a discipline. The two most common rationales are scholarly and contextual. (1) We can group various phenomena together because we as scholars intuitively see them all as part of what we mean by religion and we like the idea of continuing the conversation about how these things compare and contrast (an international scholarly rationale). In this case we as scholars of religion are making "religion" a second-order (scholarly) concept. For arguments along these lines, see Capps 1995, 336–37; Braun 2000, 14–15; and Tweed 2006, 30–33. (2) We can group them together because there is a compelling reason to do so in our local/national context. As Saler (2000, 21) points out, "a serviceable, explicit definition of religion would seem to be a practical necessity for certain public agencies in the United States." Given the importance of the legal category in the United States, it makes sense in that context, as one of my graduate students has argued, to have places—for example, departments of religion—in which to reflect on the set of things that might come under that heading.

broader, more generic ways of identifying the sorts of ascriptions that interest us. Rather than relying on emically loaded first-order terms, such as "sacred," "magical," "spiritual," "mystical," or "religious," scholars of religion can seek ways to translate the disciplinary second-order discourse of "religion," "religious," and "religions" into broader, more generic terms when designing their research. Instead of stipulating a definition for a key first-order term, such as "religious," and thus defining in advance what exactly will count for us as such, I propose—to borrow a fishing metaphor—that we cast our nets more broadly and then sort through the variety of things that our nets pull in. Identifying a broader net or set of nets that we can use—singly or in combination—to specify what it is we want to study would not only eliminate the confusion between first- and second-order use of the term "religious," it would also highlight the impossibility of uniquely specifying what is meant by the first-order terms and force us to specify what it is about this constellation of concepts that most interests us.

Specialness as a Generic Ascription

The idea of "specialness" is one broader, more generic net that captures most of what people have in mind when they refer to "sacred," "magical," "spiritual," "mystical," or "religious" and then some. We can consider specialness both behaviorally and substantively, asking if there are behaviors that tend to mark things off as special and if there are particular types of things that are more likely to be considered special than others.

The approach I am suggesting can be and indeed has been derived fairly directly from a certain reading of Durkheim (Hervieu-Léger 2000, 102–8; Pargament and Mahoney 2005, 179–98). As indicated at the beginning of the chapter, Durkheim defines *a religion* (not "religion") as "a unified system of beliefs and practices relative to sacred things, that is, things set apart and forbidden" (Durkheim 1912/1995, 44 [emphasis added]). While Durkheim is not entirely consistent, he makes a remarkably clear distinction between *religions* and the elementary phenomena that constitute them. Sacred things (things set apart), religious beliefs (beliefs about sacred things), and rites (rules for behavior in the presence of sacred things) can all exist apart from religions (Durkheim 1912/1995, 38).[4] Sacred things and beliefs and rites related to them are separable from religions and at the same time provide the fundamental raw material that people use to construct "religions."

[4] Durkheim (1912/1995, 39–44) also provides a tortured, and in my view entirely unhelpful, effort to distinguish "magic" and "religion," in so far as both involve things set apart, beliefs, and rites.

Durkheim's concept of "sacred things as things set apart and forbidden" can be used to generate a generic second-order concept of "specialness," if care is taken to avoid certain pitfalls.

- Given the many meanings attached to the sacred, some of them highly problematic, we can consider "the sacred" an emic term and refer simply to "things set apart and forbidden," where "thing" can literally mean *anything*, whether event, person, behavior, object, experience, or emotion. In contrast to Hervieu-Léger (2000, 106), who associates "the sacred" with a particular kind of experience (the experience of "the sacred"), we need to make a careful distinction between the infinite variety of things, including experiences, that people set apart and the way they respond to them.
- Things are always set apart relative to other things in a class. Setting something apart in this way marks it as special; we can refer to this process as one of "singularization" (Kopytoff 1986). In contrast to Pargament and Mahoney (2005, 182–83), however, singularities may be positive or negative, for example singularly good or singularly bad.
- We can locate "things set apart and forbidden" at one end of a continuum that runs from the ordinary to the special, with things that are so special that people set them apart and protect them with prohibitions or taboos at one extreme. In doing so we can avoid Durkheim's problematic claim that religious thought divides the world "into two domains, one containing all that is sacred and the other all that is profane" (Durkheim 1912/1995, 34) and refer simply to things that are more or less special, things understood to be singular, and things that people set apart and protect with prohibitions.

Rather than stipulating a definition of "religious ascriptions" or "things deemed religious," we can use the idea of "specialness" to identify a set of things that includes much of what people have in mind when they refer to things as "sacred," "magical," "mystical," "superstitious," "spiritual," and/or "religious." Whatever else they are, things that get caught up in the web of relations marked out by these terms are things that someone or some group has granted some sort of special status. Whether or not particular things should be considered special is typically a matter of dispute and leads different individuals and groups to position things differently in relation to the web of related concepts. Although neither the specific Western terms nor the web of relationships that encompass them correspond precisely to distinctions made in other cultural contexts, the concept of "specialness," in so far as we can operationalize it in terms of behaviors, provides a more promising starting point for cross-cultural and cross-temporal research. Such a starting point will most likely capture

more than what scholars might normally (that is, from our culturally located perspective) consider "religious," "spiritual," "magical," and so forth.
Rather than viewing this as a cause for concern, this gap will allow us
to view ourselves and what we tacitly consider "religious," "spiritual,"
"magical," et cetera in a more reflexive light.

SPECIAL THINGS AND THINGS SET APART

In what follows, I want to consider a series of questions about specialness:
first, whether there are particular behaviors that can help us to identify
things that people consider special; second, whether there are some kinds
of things that people are more likely to consider special than others; and
third, what methods are needed in order to interact with these things.

In the first part of this section on specialness, I am going to come at
the issue from a formal or functional perspective, using Durkheim's definition of sacredness as "things set apart and protected by taboos" as a
starting point. By specifying various prohibitions commonly used to set
things apart as special, we can identify formal marks of specialness that
should be usable across cultures and time periods. In the second part of
the section, I will approach the matter of specialness more substantively,
asking if we can identify more substantive characteristics of things that
tend to be set apart as special. Here I will suggest that both anomalous
and ideal things may tend to stand out for people and that much of what
we traditionally associate with religion—for example, spiritual beings and
abstract concepts such as transcendence, ultimate concern, and nirvana—
can be interpreted as either anomalous or ideal. Since the features that
characterize specialness are neither mutually exclusive nor co-extensive,
scholars will have to decide which they want to highlight in their research.
In the third part of the section, I will consider the special methods required
to interact with special things in so far as (1) they are set apart by prohibitions that preclude interaction and (2) they or their special qualities are
evanescent. To the extent people do not care to attend to special things,
such methods are not necessary. The discussion of special methods (for
example, practices, rituals, prayer, chanting, divination) for engaging with
special things will require us to distinguish between simple ascriptions
(things deemed special) and composite ascriptions (methods deemed efficacious for engaging with things deemed special).

Throughout I will reflect—albeit quite tentatively—on the question of
why humans singularize things. Distinguishing between simple and composite ascriptions of specialness allows us to identify basic building blocks
that can be incorporated into more complex socio-cultural formations.
Comparison of humans across cultures and with other animals may allow

us to consider the interplay between things that humans are biologically primed to view as special, on the one hand, and culturally primed to view as special, on the other.

In looking at the concept of specialness formally, substantively, practically, and comparatively, I am not attempting to create a new, more embracing definition of "religion." Instead I am breaking down what we have generally meant by "religion" into "religious things" and "religions" and then positioning both within a larger, more encompassing framework of special things and the methods deemed efficacious for engaging with them. Doing so will allow us to specify an object of study more precisely and set up comparisons that take these differences into account.

Marks of Specialness

Durkheim's definition does not provide a precise sense of what researchers should look for if they seek to identify things that people consider special and, in more extreme cases, set apart and protect by taboos. Drawing from classical debates over commodities (Marx, Simmel), fetishes (Marx), the sacred (Durkheim), and gifts (Mauss), scholars in various disciplines have proposed ways to operationalize these prohibitions in relation to wider discussions of where value is or ought to be placed in relation to concrete objects, relationships, and abstractions. These discussions are typically framed against the backdrop of economic exchange and identify various prohibitions as protecting forms of valuation that exist alongside or in opposition to economic exchange value.

Building on Durkheim, Igor Kopytoff (1986, 73–75) introduced the idea of "singularization" as one method that people or groups use to "set apart a certain portion of their environment, marking it as 'sacred.'" As defined by Kopytoff, singularities are things that individuals, groups, or societies refuse to commodify, implicitly marking them as priceless. Societies typically have a set of things—"a symbolic inventory"—that they consider singular, such as public lands, historic monuments, state museums, and symbols of office from government buildings to artifacts, and that they refuse to exchange. Nations, groups, and individuals do not necessarily agree on what counts as singular, however, as is evident in disputes between native peoples and the governments of the United States and Canada over lands that native peoples consider sacred or debates over what events should be memorialized and in what fashion (Chidester and Linenthal 1995).

Though some people may understand singularity to inhere in the thing itself, things often pass from one status to the other and back. Many religious traditions have prescribed methods for sacralizing (consecrating) and desacralizing (deconsecrating) objects. In the ancient Near East and in India, elaborate rituals transformed human-made statues into cult objects in which deities resided.[5] More recently, anthropologists have reported on Thais who have singularized trees by "ordaining" them (wrapping them in the saffron-colored cloth associated with Buddhist monks) and by photographing them in order to prevent their destruction (Vail 2006). Although singularities are usually highly valued, they may also be completely shunned, as in the case of things viewed as demonic or evil. Things are not necessarily completely singular or completely commodified. U.S. flags are technically commodities, but are nonetheless considered special. Flag etiquette requires that the flag should not touch the ground, should be illuminated at night, and should be destroyed respectfully if damaged. There have been numerous attempts to pass a Flag Desecration Amendment to the U.S. Constitution. Other things that are set apart may be traded within a very narrow sphere of exchange or only under certain conditions. The U.S. government allows logging in national forests under certain conditions and generally prohibits it in national parks. Whether drilling for oil should be allowed in the Arctic National Wildlife Refuge is hotly debated. Other commodities, such as papal indulgences, medicines, and certain forms of software, are intended for their original user; they can be sold once, but cannot be resold.

There are often intimate connections between singularity and identity, whether individual or collective. Anthropologist Annette Weiner (1985, 212) introduced the idea of inalienable possessions: that is, possessions that have histories that are carried in the object whether they are kept or circulated to others as gifts, loans, or copies, and, as a result, establish an "emotional lien upon the receiver." Patrick Geary (1986) has shown how medieval Christian relics, as physical remains of special (that is, saintly) dead people, were normally given as gifts, thus retaining their ties to the donor, establishing a lien on the receiver, and constituting networks of ecclesial patronage. Relics that were stolen or sold were alienated from the giver and escaped these patronage relationships.

Some scholars have questioned the value of Durkheim's definition of the sacred as things set apart, arguing that his understanding is overly dichotomous and cannot account for the ways in which sacred things are embodied materially and exchanged commercially. Colleen McDannell (1995, 4–8, 132–62), for example, has made this argument with respect to the international distribution of holy water from the Catholic shrine

---

[5] On the animation of statues, see Dirk (1999), and especially Waghorne's article (1999) on the making of cult images and the rituals that consecrate them. "The members of Sai Samaj in [present day] Madras . . . expect[] that their own actions in ritual will bring the living presence of divinity into a marble image formed by human hands" (214). She also witnessed similar rituals at temples in the United States.

at Lourdes, and Suzanne Kaufman (2005, 7–9), drawing on McDannell's critique, has done so in relation to the commercial activities surrounding the shrine itself.

Although they rightly criticize the idea of two distinct and comprehensive domains of the sacred and profane, it is clear from McDannell's account that the holy water was not actually sold. Distributors were given the water from the shrine in return for donations. Water was distributed from replicas of Lourdes in the United States and elsewhere, again in return for donations instead of payments. Lourdes water currently on sale on the Internet would seem to belie this point, but at least one site is careful to state, "We do not sell the water but sell the container to off-set import costs."[6] Although this may seem like a technical nicety, it is a technicality that matters, at least to the more devout. From a scholarly perspective, it is a subtle but crucial boundary that sets the water apart as a gift, not a commodity. As long as this boundary is honored, everything else can be commodified; representations of Bernadette, bottles for the water, souvenirs of Lourdes, package tours, et cetera, can all be sold.

The Catholic tradition, which enshrines in canon law precise rules regarding the treatment of holy objects, may lead Catholics to make distinctions that others would not. Commercial activity associated with other shrines may offer more support for the idea that specialness and commodification are not necessarily opposed. Nurit Zaidman's (2003) research on the religious objects sold at Jewish pilgrimage sites and New Age shops in Israel indicates that there are many things that people consider special, yet not so special that they cannot be bought or sold.

In the case of the shrine at Lourdes, however, most Catholics do not consider the water holy special in its own right, but by virtue of its association with an alleged appearance of the Virgin Mary to Bernadette Soubrious. The primary thing that was set apart, in other words, was not an object (the water), but an event (a series of visionary experiences). While prohibitions against selling or trading are commonly used to set objects apart, people often use other, more abstract prohibitions to set apart more abstract and ephemeral things. In his research on "sacred values," psychologist Philip Tetlock (Tetlock et al. 2000) highlights the importance of taboos against mixing and comparing for setting apart more abstract and ephemeral things. Thus, in the case of Lourdes, Bernadette's visionary experience was not set apart by means of prohibitions against selling accounts of it, which definitely were sold, but by precluding comparisons between it and other similar experiences. When the claims of competing visionaries threatened to undermine the specialness of Bernadette's experience, her backers responded by discrediting her competitors' visions

[6] See http://ourladyoflourdescatholicgifts.com/LourdesWater.html.

as inauthentic. In doing so, they reasserted the singularity of her visions and set them apart from others they deemed inauthentic (Zimdar-Swartz 1991, 57–67; Harris 1999, 91–109).

If we think more carefully about what is involved in attaching a price to something, we can see how pricing, mixing, and comparison are related. First and most basically, setting a price means specifying the value of a thing in relation to other things. If something is priceless, it cannot be compared with other things in this way because doing so reduces it to something that can be bought and sold. Indeed, the mere thought of attaching a price to things such as friends, children, or country sparks feelings of outrage in most people. This powerful emotional response to the idea of putting a price on something priceless points to a second, less rational aspect of pricing. Comparing something of infinite value to something of finite value does not simply reduce the value of the former in a purely rational economic sense. Emotionally, the reduction of a priceless thing to something that can be bought and sold degrades it in the most literal sense of the term. By extension, something of infinite value is polluted and contaminated when it is mixed with or even compared to things of finite value. While there may be some things that literally cannot be sold, anything, however abstract, is potentially subject to degradation by means of mental mixing or comparison. Prohibitions against mixing and comparison thus provide the most comprehensive means of setting things apart, whether they are objects, persons, events, experiences, or ideas. Framed positively, prohibitions against trading, mixing, and comparing allow people to set things apart as priceless, pure, and incomparable.

Where Tetlock expands the price/priceless binary to include comparable/incomparable and mixable/unmixable, Graeber (2001) suggests three alternative measures of value (proportionality, ranking, and presence/absence) that intersect with those proposed by Tetlock. Proportional value refers to value relative to other things and thus involves price, comparison, and implicit mixing with other things. Things that can be compared but not priced can be ranked. Thus, in some frames of reference, Jesus can be ranked more highly than the saints, Allah more highly than jinn, celibacy more highly than marriage, the aristocracy more highly than the bourgeoisie, Brahmins more highly than Kshatriyas, and so on. Even when things cannot be priced, mixed, or compared, Graeber points out that "there is still the difference between having them (or otherwise being identified with them) and not" (ibid., 75). So, for example, in the wake of the California Supreme Court decision allowing same-sex couples to marry, Pope Benedict XVI stated that heterosexual marriage should be set apart from all other types of unions and not compared to or confused with them. Heterosexual marriage, he said, "represents a good for all society that can not be substituted by, confused with, or compared to other types of unions" (Pullella 2008). Though the pope held up heterosexual marriage as unique and incommensurable, it remains for individuals to decide whether to hold to or identify with this value or not.

Much of the complexity that surrounds the setting apart of objects applies as well to the setting apart of things such as behaviors, experiences, and events. Thus a society's symbolic inventory is not limited to tangible places, structures, and objects, but may also include events, experiences, and ideas that may be recounted in narratives and embodied in beliefs. These more ephemeral things, like objects, may be set apart as singularities or placed on a level with other events, experiences, and behaviors. As is the case with objects, the value of the thing set apart in this way does not have to be positive; things with an infinitely negative value—for example, things that are evil or absolutely disgusting—may also be set apart. Some people, for example, set apart the Holocaust, treating it as a uniquely horrifying event that cannot be compared to other events without compromising it (Linenthal 1995).

Though singularity may seem to inhere in certain events and experiences, here, too, things often pass from one status to the other and back, whether in the minds of individuals or within or between groups. Thus, while adherents understand key events as singular (for example, the giving of the tablets to Moses on Mount Sinai, the revelation of the Quran through Mohammed, the incarnation of God in Christ, the Enlightenment of the Buddha), competing traditions and secular methods of study can and do cast them in terms that effectively desingularize them and make them more ordinary. As with singular objects, we can expect to find that other sorts of singularities are also bound up with matters of identity and that they can be transmitted (or circulated) in ways that preserve that identity more or less intact or in ways alienate them from meanings people have ascribed to them in the past. Thus people are more likely to recount events they consider singular as narratives and reenact them in rituals than to use historical-critical methods to study them alongside other similar events. Competing notions of how past events should be remembered can lead to intense controversy over how memories should be enshrined in memorials, as was the case, for example, with the Holocaust Museum in Washington, D.C. (Linenthal 1995), or the way monuments should be interpreted, as in the case of Mount Rushmore (Glass 1995).

In light of this expanded understanding of the ways things can be set apart, we can interpret the sui generis approach to religion as one that sets religion apart and protects it with taboos. In asserting that religion is sui generis, advocates of this approach assert their belief that religion is inherently special. The disciplinary axiom that prohibits "reducing" religion to something else protects religion from comparison with nonreligious things and from explanations that cast it in nonreligious terms. Viewed ascriptively, those who view religion as sui generis set it apart from other objects of study and protect it—typically without any further justification—from profanation by means of the taboo against reductionism.

When people set something apart and protect its specialness with prohibitions against selling, trading, mixing, or comparing it with ordinary things, they assume that violating those prohibitions will cause something bad to happen. Narrowly conceived, human violation of the taboo will in fact make the special thing ordinary; broadly conceived, however, it will cause everything the thing set apart represents to collapse (that is, specific relationships and, by extension, potentially the whole social and cosmic order). In the case of the sui generis approach to religion, scholars seem to fear that violating the taboo against reductionism will destroy "religion," or at least deprive it of its specialness. If religion is not special and set apart, this in turn may lead some to question the need for special departments devoted to its study and thus, by extension, threaten the "cosmic order" of the university (or at least a small corner of the humanities).

If violating taboos causes things to happen, so, too, setting things apart, as Durkheim recognized, configures social relations in certain ways. When a woman sets apart a particular infant as hers, she constitutes a mother-child bond between herself and that infant. Should she violate the taboo against exchanging her infant for any other infant, she violates the taboo that sets her infant apart from all other infants and destroys the mother-child bond between them. Violating the taboo that sets the thing apart thus does more than "profane" the thing in question; it also destroys the singular relationship between the thing and those who set it apart. The thing set apart can play an entirely passive role in this process and thus does not need to be an agent in its own right. The only agency required is that of the individual or group that sets something apart and observes the taboos surrounding it. In doing so, they both constitute the thing as special and constitute their own special relationship with the thing. If they violate the taboo associated with the special thing, they allow it to become an ordinary thing and cause the special relationship between themselves and that particular thing to disappear.

Whether we view specialness as inherent in some things or as ascribed to them, it would appear that the tendency to set some things apart as special is a deeply rooted human characteristic. Although there has not been a great deal of attention given to the biological foundations of specialness as such, some researchers (Durkheim 1912/1995; Graeber 2005, 521; Kirkpatrick 2005, 246–47; Dow 2007, 8) suggest that the inviolability of sacred things is rooted in relationships (for example, parent-child, mate, leader-follower) that are essential to the formation and stability of groups (for example, family, herd, tribe). If this is correct, it would suggest that the ability to set things apart is not simply a human characteristic but a mammalian one rooted in the mother-infant bond and extended by some mammals to relationships between mating pairs and hierarchical relations between leaders and followers (Panksepp 1998, 246–60). If, as this line of reasoning would suggest, the idea of set-apartness is implicit in relationships common to primates and other mammals, then we can assume that it is extended by analogy both to objects and linguistic abstractions that are linked in some way to relationships, whether these are basic relationships or more complex ones.

Within such a theoretical framework one would expect that people would not only view some relationships as inviolable but would also ascribe more value to objects or abstractions associated with inviolable relationships than to those with no such associations. This enhancement of value is evident in what economists refer to as the "endowment effect," which refers to the tendency of people to attribute a higher price to an object they have been given than to an identical object they could purchase in a store. In a series of experiments, researchers found that the disparity increased dramatically (from 2:1 to 8:1) when the objects in question were acquired via communal or family rituals. Moreover, when the objects were acquired from intimate partners, people were often unwilling to assign any price to them at all (Tetlock 2003, 322). Though further research on this is needed, it may be that the value of the things that people set apart in this way is derived directly or indirectly from relationships that humans as a species need to survive and which, for that reason, people viewed as inviolable.

Types of Specialness

In the discussion thus far, we have looked at specialness as a function of value, such that singular things are things whose value cannot be specified relative to other things. Viewed in this way, specialness implies the *belief* that some things are priceless or incomparable, the *ascription* of singular value to things, and the *attribution* of causality to the violation of the taboos associated with those things. Value is not the only way to think about specialness, however. As a formal feature of specialness, it tells us more about the form that specialness takes (its set-apartness), than about specialness per se. Now we need to consider whether there are some types of things that are more likely to be considered special than others. Are there characteristic features that tend to make some things stand out as special regardless of whether people consider them so special that they set them apart with prohibitions? While this is ultimately a matter for empirical research, we can further such efforts by distinguishing between two logically distinct types of singularity: ideal things and anomalous things. We can locate both on continua from the ordinary to the special and consider what happens when people consider them so special that they set them apart and protect them by prohibitions.

Ideal things. Let us begin by considering things that stand out as special because they seem ideal, perfect, or complete. They may stand out in this way in a relative sense or, if they are thought to approach an ultimate horizon or limit, they may signal an ideal in an absolute sense. As absolutes, they are no longer on a continuum with limited things, however special such things might be, but are fully set apart. Many of the qualities that we use to designate something as special (for example, true, real, good, beautiful, pure, natural, bad) can be transformed into absolutes. English-speakers often turn these qualities into absolutes by nominalizing and capitalizing them. Thus, for example, "beautiful" may be used to designate a flower as special ("a beautiful flower") or transformed into an absolute by nominalizing and capitalizing it as "Beauty." Many other terms that we use to mark things as special can be turned into absolutes in the same way, including "Transcendence," "the Infinite," "Reality," "Truth," "Good," "Evil," "Purity," "Perfection," "Nature," and "Ground of Being." The language of specialness and things set apart thus allows us to distinguish between terms that designate something as special ("beautiful") and terms that set things apart as absolutes ("Beauty").

The language of specialness and things set apart also allows us to sidestep the question of whether an absolute, such as "Beauty," is a philosophical or a religious concept in order to focus on the features that distinguish it from a quality that simply marks something as special. The chief feature of the fully set-apart absolute, I want to suggest, is its (postulated) existence apart from human perception and imagination. If we describe a flower as beautiful, we perceive a quality in the flower and mark the flower as special, in this case beautiful. If the perceived beauty of the flower evokes a response in us and we attribute the flower's ability to evoke this response to Beauty manifesting itself in the flower, then we have transformed a perceived quality into an absolute. Viewed in this way, the absolute manifests itself in things and people recognize and respond to it as something that exists apart from themselves and the things in which it is manifest. Thus, from an attributional perspective, we can say that people *ascribe* this absolute character to things, including objects, agents, and experiences. If they believe that it exists in itself, they can search for it, cultivate it, recognize it, respond to it, and attribute causality to it. People can conceive of an absolute, in other words, that exists in itself (is not created by them) and causes things to happen in a purely passive sense (without intentional agency).

Sometimes an absolute is subtly present within a seemingly nondualistic naturalism. Philosopher Owen Flanagan (2005), for example, claims to detect a subtle dualism in the Dalai Lama's nontheistic Buddhism. Flanagan notes that, although the Dalai Lama appreciates the fact that the mind has neural correlates and supports neuroscientific research on meditation, he nonetheless asserts that "there is no reason to believe that the innate mind, the very essential luminous nature of awareness, has neural correlates, because it is not physical, not contingent upon the brain." For the Dalai Lama "the very essential luminous nature of awareness" is an absolute, such that "brain and mind [at "a more subtle level of consciousness"] are two separate entities" (Tenzin Gyatso, quoted in Flanagan 2007, 87). The "innate mind" thus is set apart from "mind" in the ordinary sense and made into an absolute that is dualistically opposed to "brain." From an attributional perspective, we could say that the Dalai Lama, as characterized by Flanagan, ascribed an innate essence to the mind and assumes that with sufficient practice those well advanced in the tradition are able to experience this essential luminosity directly.

Alternatively, people may view themselves as creating something in which the absolute can manifest itself. Thus, for example, artists often create things that they and others consider beautiful. Even though everyone acknowledges that artists create works of art, artists and others may feel that the standards of beauty used to judge their work are absolute. Wassily Kandinsky, for example, characterized the artist as the "priest of beauty." Beauty was to be sought in "the principle of inner need," which springs from the soul, and allows for the "ever-advancing expression of the eternal and objective in terms of the periodic and subjective" (1911/2006, 67–68, 108). Here beauty as eternal and objective is fully set apart from the periodic and subjective even though expressed through it.

Flanagan's recent attempt to articulate a nondualistic, nontheistic spirituality oriented toward good, truth, and beauty rejects what I am calling fully set-apart absolutes in favor of absolutes that are not completely set apart from the ordinary, natural world. While Plato's unchanging Forms (for example, the Good, the True, and the Beautiful) are fully set apart from the flux and change of the material world, Flanagan recasts the Platonic ideals—in light of his scientific, evolutionary perspective—as natural goals toward which humans orient themselves (Flanagan 2007, 39–44, 187). As natural goals, they are not fully set apart from the natural, but nonetheless still partake functionally of the character of absolutes by virtue of a human desire for transcendence and a human ability to imagine absolutes. As such, he roots a human orientation toward these goals in a naturalistic evolutionary perspective compatible with the ascriptive approach I am taking.

To be worth much, however, an ascriptive approach must be able to account for both the feeling that a quality of absoluteness inheres in a things and the sense that created things may reflect or embody an absolute. Flanagan (2007, 197–98) suggests that these ideals are the outgrowth of a natural disposition to make sense of things, which he roots in the evolution of the moral emotions. Though still at a preliminary stage, research on emotions, both the primary emotions common to all animals and the more elaborated moral emotions shared to some degree by primates and humans, is potentially suggestive in this regard. Thus, for example, Jonathan Haidt and others (Haidt 2003, 2007) upend the Platonic view of reason as ruler of the passions and identify five affective foundations, each with separate evolutionary origins, upon which they believe human cultures construct moral communities. These psychological foundations, they argue, give rise to intuitions having to do with harm and fairness; in-group/out-group dynamics and loyalty; social hierarchy, authority, and deference; and bodily defilement and purity. Their research suggests that all of these moral intuitions can be caught up in culturally divergent processes of meaning making and formulated as absolutes.

Haidt and Keltner (Haidt 2000; Keltner and Haidt 2003) have also argued for a family of awe-related emotions, which they suggest are evoked by a sense of vastness that cannot readily be accommodated within existing frames of meaning. Vastness can take various forms and can be elicited by powerful individuals, nature, and art. In a more recent study, Shiota, Keltner, and Mossman (2007) found that awe is elicited by a heterogeneous set of experiences, the most common of which are experiences of natural beauty, artistic beauty, and exemplary or exceptional human actions or abilities.

Although Flanagan attempts to locate a capacity for transcendence, defined as a latent capacity for meaning making, at the level of the moral emotions, it may be that the urge to make sense of things is rooted more basically than that. Focusing on primary rather than moral emotions, neuropsychologists Jaak Panksepp and Douglas Watt (Panksepp 1998; Watt 2007, 121) identify two basic clusters of emotions (a social-connection cluster and an organism-defense cluster), as well as a "seeking system" that drives organisms to engage with other living things in order to meet their essential emotional and biological needs. Flanagan's notion of transcendence might have its biological roots in something like Panksepp's "seeking system" (Watt 2005; Panksepp 2007).

Anomalous things. Anomalies are things that people consider special because they are strange, unusual, or in some way violate people's expectations. Anything can be anomalous, but we will be primarily concerned with events, natural objects and landscapes, and experiences (feelings, sensations, and perceptions). Anomalies can include unusual natural events (for example, comets, earthquakes, eclipses, and solar auras) and things with an unusual appearance (for example, "monstrous" births, faces in clouds, lifelike rock formations) (Wittkower 1942; Daston 1991; Shrady 2008). Shanafelt (2004, 336) places all such occurrences under the heading of "marvels"—that is "event[s] or effect[s] of extraordinary wonder, thought to be tangibly real, that [are] claimed to be the result of ultra-natural force."

A few researchers have focused specifically on unusual experiences and we can use their lists to suggest some of the sorts of experiences that may be widely considered as anomalous. Under the heading of "wondrous events," McClenon (1994, 1) places phenomena thought to exceed scientific explanation, such as extrasensory perception, apparitions, out-of-body and near-death experiences, spiritual possession, immunity to pain and heat, psychokinesis ("mind over matter"), poltergeists, miraculous healing, and contact with the dead. Psychologists Cardeña, Lynn, and Krippner (2000) use the term "anomalous experience" to refer to hallucinations, synaesthesia, lucid dreaming, out-of-body and near-death experiences, psi-related (parapsychological) experiences, alien abduction, past-life awareness, unusual healing, and mystical experiences.

Whether people consider something anomalous may be highly contextually dependent or it may not. With naturally occurring events, such as hurricanes or earthquakes, geography plays a part in shaping what people consider anomalous. What might seem anomalous to people in California (hurricanes) would not seem at all strange to people in Florida. Some experiences, such as spirit possession, may be considered strange or unusual in one culture and not in another. In chapter 4, we will discuss ways of determining the cross-cultural stability of various unusual experiences regardless of whether they are considered anomalous or not. For now, I want to focus on the implications of the distinction between special things and things so special that they are set apart and protected by taboos. As with our discussion of ideal things, we can consider anomalous things as special, locating them somewhere on a continuum from ordinary to very special, without necessarily considering them so special that they are set apart with taboos against comparison and the like.

In discussing anomalous things, we can distinguish between anomalous events, places, objects, and experiences that do not suggest the presence of anomalous agents and anomalous experiences of agency (that is, perceptions, sensations, or feelings that do suggest the presence of an unusual agent. Let us consider anomalous things that do not suggest the presence of unusual agency first, as they are similar in many ways to ideal things. Here, as in the case of ideal things, we have things—whether events, places, objects, or experiences—with anomalous characteristics that cause them to stand out and make it more likely that people will consider them special. Just as qualities associated with ideal things can be absolutized and set apart, so, too, qualities associated with anomalous things can be reified and set apart.

People often use the terms "mystical" and "spiritual" to mark events, places, objects, or experiences as very special and, depending on what they mean by the terms, sometimes to signal that they consider the thing in question so special that it cannot or should not be compared to other, less special things. People may use these terms to mark things as belonging to another realm or manifesting a different sort of energy or exemplifying a higher aspect of reality that is not just special, but so special that it cannot be compared with more ordinary things.

We can take as an example the anomalous experience in which the boundary between self and world seems to dissolve. Most people would probably consider such an experience unusual and thus special to some degree. Some philosophers of religion, as we have seen, characterize such experiences as mystical and in so doing mark them as very special. In chapter 2, we will discuss the experience of "pure consciousness," which Robert Forman characterizes as mystical, and, in chapter 3, William Barnard's experience of "a surging, ecstatic, boundless state of consciousness," which he, too, describes as mystical. The use of the term "mystical" in all these cases sets the experience apart from other experiences, tacitly asserting that it cannot or should not be compared with experiences that are not viewed in that way.

Seen in this way, the mystical and spiritual are treated as qualities that manifest themselves in (anomalous) things; people recognize and respond to these qualities as to something that exists apart from themselves, even when they are manifest through their own bodies. Thus, from an attributional perspective, we can say that people *ascribe* this mystical or spiritual character to things, including objects, places, events, and experiences. In so far as they are not considered agents, neither the mystical nor the spiritual acts intentionally, but in so far as people believe that these qualities can be manifest in themselves and other things, they can search for them, cultivate them, recognize them, respond to them, and attribute causality to them in a passive sense, without viewing themselves as creating them. When they respond to such qualities, people believe they respond because the qualities exist as such independent of them and have the power to passively elicit a response in them.

Although anomalous events, places, objects, and experiences characterized as mystical or spiritual are not experiences *of* an agent, people may still attribute them *to* an agent, if they believe that there are agents who can and do cause such things to occur. We can distinguish, in other words, between things that are attributed to the action of *an agent*—that is, things people believe were caused by an agent—and feelings, sensations, and perceptions *suggestive of agency*. In the first instance, an agent is presumed to exist and something is attributed to the presumed agent based on people's knowledge of or beliefs about the agent in question. In the second instance, we have feelings, sensations, and perceptions that are suggestive of agency. On the basis of this type of experience and the beliefs they hold about it, people may postulate the presence of an agent to whom they then can attribute the power to act and affect things.

We can extend this distinction between attributions to agents and suggestions of agency to make a parallel distinction between postulated *anomalous* agents and feelings, sensations, and perceptions suggestive of *anomalous* agency. Claims regarding the actions of anomalous agents presuppose their existence and are based on what people believe such agents are likely to do. Claims of the second sort can and do exist apart from beliefs about the existence of anomalous beings. Beliefs about anomalous agents often play a considerable role in determining whether experiences suggestive of anomalous agency are transformed into claims about the presence of anomalous agents.

The movement from *suggestions of anomalous agency* to *postulated anomalous agents* is a key step on the continuum of specialness. Thus, anomalous experiences suggestive of agency are likely to be considered special, but not exceedingly so as long as the suggestions of agency (for example, the sense of an unseen presence in the room) are not taken to be signs of the actual presence of an anomalous agent. We will consider the anomalous experience of a "felt presence" in detail in chapter 4. For now, it is enough to recognize that most people would consider a vivid experience of an unseen "felt presence" as unusual and thus to some degree special. Some philosophers of religion, as we have seen, characterize such experiences as numinous and in so doing mark them as very special. Although some people might view the experience as (say) a hallucination, we can assume that in many cases the *feeling* of presence (the intimations of agency) will be attributed to the *actual* presence of an invisible agent. In such cases, the assumed presence of a real entity that is capable of performing intended actions would lead most people to consider the experience more unusual and, thus, more special. When people interpret feelings, perceptions, or sensations that are suggestive of agency as evidence of the presence of an actual agent, they attribute the experience to an external source. Thus, from an attributional perspective, we can say that people *ascribe* intentionality to things, thus creating agents; they then can attribute *causality* to the agent, assuming that the agent *caused* the experience in question.

People will in turn consider signs of the actual presence of an anomalous agent more or less special depending on the beliefs they hold about them. If people believe in the existence of anomalous agents, such as spirits, ancestors, ghosts, demons, fairies, and deities, then the specialness of signs of the actual presence of an anomalous agent will depend on which agent is presumed to be present and how special they consider that agent to be. People clearly do not consider all anomalous agents so special that they set them apart with taboos. If the unseen presence were thought to be a ghost, people would not consider it nearly as special as if it were thought to be a divinity.

Responses to Bernadette's vision of Mary illustrate the role of beliefs about anomalous agents in determining how special people consider an event to be. Devout Catholics considered the apparition to Bernadette very special not only because it was Mary who appeared but also because Mary referred to herself as the "Immaculate Conception." In doing so, the alleged apparition confirmed Pope Pius IX's definition of the doctrine of the Immaculate Conception, promulgated in 1854, and contributed to the general mystique (or aura of specialness) surrounding the papacy in the run up to the declaration of papal infallibility in 1870 (Zimdar-Swartz 1991, 55–57; Kselman 1992, 92–94). Skeptics, however, viewed the apparition as a hallucination. In the eyes of the latter it was still anomalous, in the sense that waking hallucinations are somewhat unusual, but not particularly special.

> *Excursus on spiritual beings as anomalous agents.* Since spiritual beings are the defining feature of religion for many people (Spiro 1966, 94; Steadman, Palmer, and Tilley 1996), let us pause for a moment to consider them in more detail. Specifically, what does it mean to refer to spiritual beings as "anomalous agents"? For whom, exactly, are they anomalous? For those who believe in them and relate to them on a regular basis? For those who do not believe in them? Or is there some more universal sense in which they might be considered anomalies?
>
> Cognitive scientists of religion, who have focused attention on beings of this sort, use the concept of "counterintuitiveness" to locate what is anomalous about them in (presumably) panhuman cognitive processes (Boyer 2001; Barrett 2004). This concept is premised on the discovery that regardless of culture people divide things into the basic categories of persons, animals, plants, natural objects, and tools. A concept is counterintuitive when it includes features that violate the characteristics that humans normally associate with these basic categories (Boyer 2001, 65). Counterintuitive agents, in this view, are constructed through the reshuffling of attributes in ways that modestly violate what we are naturally prepared to expect.
>
> Cognitive scientists of religion argue that there is a fairly short list of violations of this sort that people have tended to view as culturally significant. Thus, Pascal Boyer (2001, 78) proposes a list that includes persons or ani-

mals that are assigned counterintuitive physical properties (for example, beings with minds but no bodies), counterintuitive biological properties (for example, beings that neither grow nor die), and counterintuitive psychological properties (for example, beings that can read minds or see into the future). Human-made objects (tools) can be assigned biological properties (for example, statues can be animated, bleed, or cry) or psychological properties (for example, the ability to hear or see).[7]

In his refinement of Boyer, Barrett (2008) identifies five ontological categories based on current research in cognitive development: spatial entities (for example, clouds), solid objects, living things that do not appear to be self-propelled (for example, plants), animates, or persons. These things can become counterintuitive agents in one of two ways: agents (that is, persons and animates) can acquire some counterintuitive property (minds without bodies, omniscience, et cetera), or nonagents (that is, spatial entities, solid objects, and things that do not appear to be self-propelled) can acquire attributes of humans or other animates (for example, the ability to talk, feel, move, think) and thus the ability to act as agents. The first subgroup, agents that acquire exceptional properties, includes the usual religious suspects—deities, spirits, demons, ancestors, souls of the dead, ghosts, fairies. The second subgroup, nonagents that acquire attributes of humans or other animates, would include, for example, consecrated statues of deities, a crucifix that moves, a statue that weeps or bleeds, and objects that are believed to contain a spirit or carry a blessing (Christian 1992; Dirk 1999; Zaidman 2003).

Scholars have developed several theories to explain the widespread cross-cultural tendency to postulate the existence of anomalous (or, more technically, counterintuitive) agents, all of which have been derived from experimental research into the workings of the human mind as viewed in an evolutionary perspective. Of these theories, which are complementary, two are cognitive and the third is affective. The first cognitive explanation theorizes that human beings ascribe counterintuitive agent-related properties to objects because they have a basic tendency to overattribute agency, particularly in situations of ambiguity. These theorists speculate that this tendency to attribute agency so capaciously has evolutionary adaptive value. When leaves rustle, for example, it is a better bet to assume that an unseen preda-

---

[7] Barrett (2008) reviews the empirical support for Boyer's hypothesis that concepts with a small number of counterintuitive features are more easily remembered and more faithfully communicated than either extremely counterintuitive concepts or more ordinary ones. Though the empirical support was mixed, he concludes that differences in how researchers operationalized "counterintuitive" might account for the results. He presents a formal system for coding and quantifying the "counterintuitiveness" of a concept that should be consulted by those who want to use the concept for research purposes.

tor is lurking than to assume it is just the wind and thus risk getting eaten (Boyer 2001, 137–67; Atran and Norenzayan 2004; Barrett 2004).

The second cognitive explanation, which emerges out of developmental psychology, suggests that humans are readily able to conceive of agents (spirits, souls, deities, ancestors, ghosts) without bodies or agents who inhabit others' bodies because we naturally distinguish between objects (bodies) and persons. Whatever a consistent metaphysics might suggest, the default mode for human beings, according to this research, is an intuitive or commonsense dualism grounded in two parallel, weakly integrated cognitive systems for perceiving bodies (objects) and persons (Bloom 2004; Cohen, 2008). These cognitive characteristics, they theorize, predispose people to postulate counterintuitive agents, especially in contexts in which ambiguous stimuli and cultural presuppositions would support such conclusions.

These cognitive dispositions can be grounded in turn in affective dynamics, including attachment processes, kinship relations, and hierarchical relations of dominance and subordination that are common to most primates (Kirkpatrick 2005). Veneration of ancestors, which is found is nearly all cultures (Steadman, Palmer, and Tilley 1996), may build on a combination of cognitive and affective processes to extend the attachments that individuals have with parents or other close kin who are believed to live on as unseen agents after their death (Kirkpatrick 2005, 247–48).

We can sum up this discussion of special things diagrammatically. Figure 1.1 represents the set of all special things. We have discussed two kinds of things that are often considered special (ideal things and anomalous things) and have distinguished between anomalous agents and anomalous things that lack agency. In each case, we suggested that these special things fall along continua of specialness ranging from the ordinary to the very special with some things considered so special that they are set apart and protected by taboos. In the figure, spiritual beings run the gamut from moderately to very special, with deities offered as examples of anomalous agents that may be considered so special that they are set apart and protected by taboos; absolutes are offered as a generic descriptor of ideal things that tend to be set apart, and "mystical" and "spiritual" as qualities that can be ascribed to anomalous things lacking agency in order to set them apart. The potential overlap between these things in practice is represented schematically by the overlap between the circles, such that, for example, deities (agents) with ideal properties and mystical characteristics are represented at the center of the diagram.

In discussing special things, we also considered ascriptions of characteristics and attributions of causality in relation to different types of belief (table 1.2). As we will see in chapter 3, people give different kinds of explanations for events they view as intended or unintended. Agents may cause things to happen intentionally or unintentionally; nonagents, lacking in-

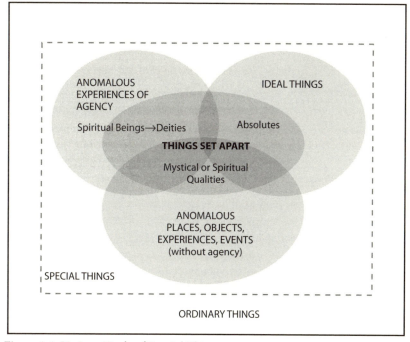

Figure 1.1. Various Kinds of Special Things

tentionality, may cause things to happen passively. In so far as singularities are not agents, people may attribute passive causality to them. Humans, whether they observe or violate the taboos surrounding things that are set apart, may do so intentionally or unintentionally. This is also the case with absolute ideals and spiritual-mystical qualities. People can characterize these ideals or qualities as existing apart from matter, but in so far as they are not agents, they do not have the power to act intentionally and thus only affect things passively. They can be manifest, exposed, or revealed passively or as a result of human action and their discovery may lead people to act in new ways. If these spiritual-mystical qualities are conceived more as energies, then they become more agent-like and, as in the case of spiritual beings, they might be understood to act with intention.

Whether people ascribe specialness to things consciously or unconsciously, the specialness of the thing is usually evanescent unless steps are taken to preserve it. The absolute may manifest in something but only those who are there will know it. A spiritual being may cause an event, but again only those who are there will witness it. If the thing or event is deemed so special that it is set apart and protected by taboos, its set-apartness makes it impossible for people to engage with it in ordinary

TABLE 1.2
Beliefs, Ascriptions, and Attributions Related to Special Things

<table>
<tr><th></th><th>Type of Belief</th><th>Ascription</th><th>Attribution</th></tr>
<tr><td rowspan="3">Kinds of Special Things</td><td>People believe some things are PRICELESS OR INCOMPARABLE</td><td>People ascribe singular VALUE to things</td><td>People may attribute passive (intentionless) causality to non-agents. Humans may violate taboos intentionally or unintentionally.</td></tr>
<tr><td>People believe in ABSOLUTE IDEALS or SPIRITUAL-MYSTICAL QUALITIES that lack agency</td><td>People ascribe EXISTENCE to these ideals or qualities apart from matter</td><td>People may attribute passive (intentionless) causality to non-agents.</td></tr>
<tr><td>People believe in the existence of SPIRITUAL BEINGS or ENERGIES that do things</td><td>People ascribe INTENTIONALITY to anomalous experiences of agency</td><td>People attribute intentional actions to anomalous agents, for which they assume the agents have reasons.</td></tr>
</table>

ways. Something more is needed if people want to engage with these evanescent and/or set apart things on a more continuous basis.

Ways of Engaging Special Things: Simple versus Composite Ascriptions

In the discussion so far, we have been envisioning special things as elementary phenomena standing on their own. Individuals or groups, however, can also incorporate them into more elaborate formations that provide means for people to engage them on a more regular or continuous basis. These formations in effect institutionalize the special things and in doing so take on the formal characteristics that in some contexts we associate with religions or spiritualities. Simple ascriptions of specialness as such are thus not *religions* or *spiritualities*, but rather the basic building blocks that people use to construct them.

Without implying that this is the only way that simple ascriptions are taken up into more complex formations, I want to suggest that the *path* is a basic concept that individuals and groups routinely use to transform religious ascriptions into protoreligions and protospiritualities. Scholars from disparate disciplines, including psychologists Pargament and Mahony and Buddhist scholars Robert Buswell and Robert Gimello, have suggested its utility in thinking about religions. Pargament and Mahony (2005, 181) define religion as a "search for significance in ways related to the sacred [such

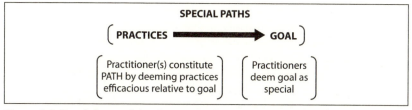

Figure 1.2. Breakdown of the Composite Ascription: Special Paths

that] . . . [e]very search consists of a pathway and a destination. Pathways are made up of beliefs, practices, relationships, and experiences that lead individuals toward their goals of greatest value." Buswell and Gimello offer a similar definition based on the Buddhist concept of *mārga*, a Sanskrit term often translated as "path." *Mārga* theory is the "theory according to which certain methods of practice, certain prescribed patterns of religious behavior, have transformative power and will lead, somewhat necessarily, to specific religious goals" (Buswell and Gimello 1992, 2–3). Drawing on the path concept, we can distinguish between *special things* and *special paths*, where special paths are defined as sets of practices that individuals or groups view as effective in attaining goals associated with special things.

The idea of a path implies both a goal and a means of getting to the goal. In contrast to the simple ascriptions already discussed, the idea of the path contains two ascriptions, one associated with the goal and the other with the means of getting to the goal. Practitioners deem the goal toward which the path aims as special in one or more of the basic ways already discussed—that is, by association with things set apart, spiritual beings, or limitless absolutes. For a path to be a path, however, practitioners must also agree on *means* that they consider efficacious for getting from where there are to the goal. We can characterize the means of getting to the goal—the things people can do to get there—as practices. The distinctive feature of a path is the linkage between the practices and the goal; this linkage is constituted when people ascribe efficacy to practices relative to a goal. When the goal of a path has to do with special things, the path incorporates the simple ascription into a composite formation made up of two ascriptions. The first ascription deems the path or, more specifically the goal toward which the path leads, as special. The second ascribes efficacy to practices in relation to the goal (figure 1.2).

The linkage that the practices constitute between the starting point and the goal may vary depending on the way the starting point is conceived, the nature of the special thing, and the way that practitioners want to engage it. If the special thing is set apart and protected by prohibitions, the practices will need to allow for engagement while at the same time maintaining the prohibitions that set the thing apart as special. If the special thing is an event associated with one or more spiritual beings, the practices may re-create the event or otherwise allow for re-engagement with the spiritual being. If the special thing is an absolute, the practices may prepare things (humans, animals, places, objects, et cetera) to manifest the absolute. However the practices constitute the linkage between the point of origin and the goal, practitioners must agree that the practices are capable of doing so—that is, that they are efficacious.

The path concept, with its source-path-goal structure, highlights the explicitly stated goals of practitioners and the means they view as effective for achieving them. It thus provides us with a basic schema for analyzing more complex behaviors from the point of view of practitioners. We can contrast it with definitions that highlight indirect effects, which subjects most likely would not view as the goal or purpose of their behavior. Thus, for example, when Durkheim (1912/1995, 44) defines "a religion" as "a unified system of beliefs and practices . . . which unite into one single moral community called a Church, all who adhere to them," he identifies an indirect effect of performing practices relative to sacred things. While this is an important effect, this is most likely not the reason that practitioners themselves would offer for why they perform the practices in question. They would most likely describe their practices as achieving something (some goal) relative to the things they believe are special.

Hervieu-Léger (2000) inverts Durkheim on this point. Rather than conceiving of beliefs and practices as things that unite people into a moral community, she suggests that groups take on a more complex form "when the group finds it needs to acquire a representation of itself that can incorporate the idea of its continuity beyond the immediate context of its members' interrelating" (151–52). Religions, in Hervieu-Léger's conception, involve chains of memory that establish continuity beyond the immediate face-to-face engagement with "the external force" that she associates with the experience of the sacred (106–7). In the terms I am using, this would suggest that people incorporate special things into more complex formations involving practices (and chains of memory related to those practices) in order to (re-)establish or maintain a connection with the special thing over time. If individuals or groups are willing to let the memory of the special thing fade without seeking to re-engage it, they most likely would have no need to incorporate it into more complex formations.

Setting up Research

I have suggested that the basic tendency to singularize is rooted in the special relationships that humans and perhaps some other animals need to survive. If that is the case, we may be biologically primed to see some things as special. Obviously, however, humans have elaborated this tendency in complex ways, infusing this biological priming with socio-cultural particularities that may enhance or undercut them. Thus we are quite capable of deciding upon reflection that something that seems special is not really so special after all and to experience something as special for us, even if it is not for others. Indeed, the very idea of specialness suggests that what each of us considers special (for example, my child, my country) strikes us as inherently more special than the things others consider special, even if we know rationally that that is not the case.

The tendency to singularize may be primal, but it is not limited to small-scale societies. Both small-scale and complex societies manifest tensions between special things and commodities and incorporate special things into more elaborate formations so they can be engaged on a regular basis. The crucial difference between the two types of societies, according to Kopytoff (1986, 79–80), is the vast proliferation of competing schemes of valuation and singularization in complex societies. The challenge for researchers, particularly in studying more complex societies, thus lies in deciding how to cut into this vast proliferation of competing schemes in order to further their own research agenda. Disaggregating "religion" into special things and special paths, each of which can be specified in particular ways, is intended as a strategy that will allow us to design research projects that probe the innumerable competing schemes of valuation and singularization in complex societies in more precise ways and, hopefully, with more cumulative effect. The various ways in which people singularize things and under some circumstances incorporate them into more elaborate formations give researchers many options when setting up an object of study. Given these more refined options for setting up research, we can examine the role of experience under different conditions and make predictions about its role that could be tested.

Constructing Objects of Study Using Simple Ascriptions

Scholars interested in simple ascriptions of specialness can set up studies that look at the process of singularization as a whole or focus on particular kinds of singularities. To study the process of singularization, researchers need to select a research site where the process is occurring. Studies by Mihaly Csikszentmihalyi and Eugene Rochberg-Halton (1981) and Nurit Zaidman (2003) focus on two different kinds of sites: the suburban home in the United States and religious sites in Israel. In their study of the meaning of things in people's homes, Csikszentmihalyi and Rochberg-Halton (1981) simply asked people in suburban Chicago to identify what objects were "special" to them and then classified objects and meanings based on their responses. The detailed information on the procedures they followed, provided in their appendices (1981, 250–89), could be used to replicate their study in other contexts. In order to compare the way objects were singularized in traditional and new religious groups, Zaidman looked at "powerful objects" sold at traditional Jewish pilgrimages sites and New Age shops in Israel. Using a similarly open-ended method of investigation, Zaidman (2003, 348–49) conducted preliminary interviews to determine what terms each group used for objects that contain power, learning that the more traditional North African Jews referred to "sacred objects," while the New Age practitioners preferred "mystic objects." In order to determine which objects were "considered to be 'religious,' 'sacred,' or 'powerful,' and what objects [did] not contain any of these meanings," interviewers then gave participants a lists of objects commonly available for sale at either the New Age shops or the shops at pilgrimage sites and asked them to group them into categories and explain their rationale for doing so.

Researchers can also set up studies that focus on particular kinds of singularities in relation to behaviors that are not necessarily singularized. Let us consider two disparate examples: exchange relationships and rituals. Neither exchange relationships nor ritualized behaviors are necessarily associated with singularities and both can be studied in relation to various types of special things. Studies by sociologist of religion Rodney Stark (1999) and psychologist Philip Tetlock (2000, 2003) illustrate how exchange relationships can be configured in relation to the basic types of religious ascriptions discussed so far. Stark looks at exchange relations as events, focusing on the exchanges themselves and, by extension, the partners to the exchanges. He defines religion in terms of supernatural beings in order to analyze exchanges between humans and gods within the wider context of exchange relations more generally. Tetlock, by contrast, focuses on the *objects* that are exchanged (or, in this case, not exchanged) rather than on the exchange *partners*. He defines religion not in terms of superhuman agents but rather in terms of sacred values (absolutes) in order to examine what kinds of things are and are not exchanged. While Stark's experiments were designed to explore the nature of exchanges between humans and postulated deities, Tetlock's were designed to explore what different communities would and would not exchange under various conditions (Tetlock et al. 2000).

Researchers can also bring both types of ascription to bear on rituals. In their theory of ritual action, McCauley and Lawson (1990, 2002) define religion in terms of supernatural agents with minimally counterintuitive properties. These counterintuitive properties, they argue, make the roles that the supernatural agents play in rituals distinct, and they use this distinction as a criterion for sorting between religious and nonreligious rituals. In his study of Japanese culture, Timothy Fitzgerald (2000a) found religion defined in terms of superhuman agents analytically irrelevant. The concepts of hierarchy, purity, pollution, and ritual proved much more analytically useful and translated easily between English and Japanese. Unlike the concept of superhuman agents, the concepts of purity and pollution, he argued, were "fundamental, and in that sense 'sacred'" to Japanese society (197).

Constructing Objects of Study Using Composite Formations

Researchers can design projects of various sorts using composite formations such as "special paths" by specifying the components (in this case, practice—goal—special) in different ways. In psychology, researchers are using this sort of formulation to investigate the relationship between personal goals and quality of life. Thus, for example, Kenneth Pargament's definition of religion as "a search for significance in ways related to the sacred" sets up an object of study in the form of a special path (Pargament 1997; Pargament, Magyar-Russell, and Murray-Swank 2005). Here "searching" is the practice, "significance" is the goal, and specialness is specified in terms of "higher powers." He and his co-authors recognize that searching is a generic phenomena, explicitly acknowledging that searches for significance are not necessarily religious and specifying that it is the involvement of the sacred in the search for significance that makes it religious. They define the sacred not in terms of an object set apart, but rather in terms of "higher powers, such as the divine, God, and the transcendent" (668). In a similar vein, psychologist Robert Emmons (2005) advocates a "goals approach" as a means of capturing the dynamic aspect of religion in people's lives and frames his research in relation to striving (a practice) for religious and spiritual goals, defined in terms of ultimate purpose (an absolute), commitment to a higher power (an abstract spiritual being), and seeking the divine in daily experience (736).

While formulations such as these allow researchers to investigate the personal spirituality of individuals in pluralistic contexts, other specifications of the components allow researchers to investigate processes within highly elaborated traditions or families of traditions. Thus, scholars of religion might focus on rituals of consecration and invocation (practices) intended to invoke the presence of a deity (a goal associated with a spiritual being). Scholars could explore this ethnographically in relation to a particular branch of a tradition, through a close reading of liturgical texts governing a particular rite, or comparatively across an array of traditions or subtraditions.

Robert Sharf (2005), for example, provides a close reading of a text describing the medieval Chinese (Chan) Buddhist "Ascending the Hall Ceremony," a monastic ritual (practice) in which the abbot delivers a formal public lecture from a ceremonial chair placed where one would normally expect to see an icon of the Buddha. In the ritual, "the abbot ascended the altar, assumed the physical posture of a buddha image, and spoke with

the authority of an enlightened patriarch" (263). The goal of the practice
was to allow the monks and visiting patrons to come "face-to-face with
a living Buddha" (265). Similarly, historians could analyze Reformation
era debates over the nature of Christ's presence in the bread and wine of
the Eucharist. Although most of the parties to these debates agreed that
the words of institution spoken in the Eucharistic ritual (practice) were
linked to the manifestation of the real presence of Christ (goal), they disagreed over the meaning of the real presence (that is, the sense in which
Christ was present) and the means whereby the practice effected the goal
(and thus in what sense the practice could be deemed efficacious) (Pelikan
1984, 187–203, 257–58).

Experience in Composite Formations

The incorporation of two ascriptions into a composite formation such
as "special path" has important consequences that are not present when
things are singularized on their own. First, the two points at which ascriptions are made means that practitioners may agree at one point but not
the other. As with the sixteenth-century Eucharistic debates, practitioners
may agree on the goal (the presence of Christ) without agreeing on how
the goal should be understood (for example, materially or spiritually) or
on how the goal can be effected (that is, on the nature of sacramental
efficacy). They may agree, in other words, on the goal, while disagreeing
with respect to the practices that constitute the path and thus most deeply
over what constitutes efficacy in relation to the goal. Other participants
to a controversy may reject the goal entirely, either redirecting the practice to another end or rejecting it altogether.

Although a great deal more can be said about such disagreements, the
most important effect of the double ascription appears in those cases where
participants collectively agree on both the goal and how it can be attained.
Where such agreement exists, a relatively closed, self-authenticating system
is created, such that, within the confines of the system, practices deemed
efficacious by the group lead to the specific religious goals and the goals
are, by definition, realized when the practices are enacted.

Within such relatively closed, self-authenticating systems, experience
takes on a different character. If the goal is formulated in experiential
terms—for example, coming face to face with a living Buddha or experiencing the real presence of Christ in the Eucharist—those who carry out
the requisite practices are understood to experience the goal in a formal
sense, regardless of what they subjectively sense or feel. Thus, Sharf's argument suggests that that *Chan* enlightenment should not be understood
solely as a subjective event occasioned through meditation nor simply as
a discursive event, but rather as "constituted in and through" rituals such

as the "Ascending the Hall Ceremony." This ceremony (a ritual practice) presupposes the goal (incarnating a fully enlightened buddha in the person of the abbot), defines what it means to realize the goal (and thus to experience it), and provides a procedure for doing so. Here, experience (enlightenment) is mediated by and subordinated to a goal-directed ritual (Sharf 2005, 257–67). The same could be said of the experience of the Real Presence in the context of the Catholic Mass.

CONCLUSION: A FOUR-FOLD MATRIX

This analysis allows us to construct a four-fold matrix (table 1.3) showing the possible relationships between the unit making the ascription(s) (individuals versus groups) and the complexity of the ascription (simple versus composite). Either individuals or groups may make simple ascriptions. Simple ascriptions always apply to a particular instance. They assert

TABLE 1.3
Variations in the Nature of Experience by Ascriptive Unit and Type of Ascription

		Ascriptive Unit	
		Individuals	*Group*
Type of Ascription	Simple	• Individuals deem a particular thing or event as special. • The experience of individuals in contact with the thing or present at the event counts as evidence for ascriptions, though others may not consider it plausible.	• Groups constitute themselves as such by reaching interpretive consensus regarding a particular thing or event. • The experience of those in contact with the thing or present at the event (witnesses, disciples) counts as evidence for ascriptions.
	Composite	• Individuals perpetuate an initial thing or event deemed special by re-creating it through practices they deem efficacious. • Individual experience counts as evidence for the efficacy of practices relative to the goal, though others may not consider it plausible.	• Groups perpetuate an initial thing or event deemed special by agreeing on how it can be re-created. • The re-creation of thing/event rests on group consensus regarding the efficacy of practices relative to the special goal, which outsiders typically do not find convincing.

that *this* thing (for example, Siddhartha Gautama's experience under the bodhi tree or Jesus' crucifixion) should be considered special and set apart, but they say nothing about how it can be recreated. When individuals or groups make simple ascriptions, they typically rely on their own experience or the experience-based testimony of others who were in contact with the object or present at the event as evidence for designating it as special. The evidence may be recent or old (for example, textual accounts of the Buddha's enlightenment or of Jesus' resurrection) depending on when the event or contact was made. The evidence may be based primarily on perceptions or feelings or a combination of both.

When individuals or groups make composite ascriptions regarding paths they set apart as special, they not only deem the thing special but also constitute a path whereby the special thing (for example, the enlightenment of the Buddha or the presence of Christ) may be re-created, re-encountered, and thus re-experienced in the present. Individuals and groups often ascribe efficacy to practices on non-experiential grounds (for example, on the authority of texts, traditions, unbroken lines of transmission).

In light of this analysis, we can ask how experience is involved in how people deem things as special. If for the moment we define experiences as things that people "have, enjoy, or undergo" (Bennett and Hacker 2003, 265), we can make the following predictions:

1. Claims regarding experience will typically play a large role in simple ascriptions. These claims may be based on new or old evidence. Emotions, sensations, or perceptions that people consider ideal or anomalous are more likely to stand out as special and to stimulate processes of reflection and explanation. These processes may in turn stimulate more elaborate ascriptions and attributions, such that in some cases people may consider an experience so special that they surround it with prohibitions or taboos.

2. In the case of composite ascriptions where a group has reached a high level of agreement on matters of ascription, we would expect to find the feelings and perceptions of subjects subordinated to the group consensus regarding what it means to experience the goal. In relatively closed, self-authenticating systems, groups do not rely on the direct perceptions or feelings of practitioners as evidence to support the underlying ascriptions. Instead, the group consensus regarding practices and goals defines and constitutes what it means to experience the goal. The subordination of individual perceptions and feelings to an agreed-upon goal and means of achieving it reduces—perhaps even eliminates—the need for direct experience of the goal in any sense recognizable to outsiders, minimizes interpretive ambiguity within the group, and ensures that experiential goals are automatically (formally) realized through the practice.

3. In groups where the level of disagreement is high relative to one or more aspects of a composite ascription, we would expect competing claimants to draw more frequently on perceptions or feelings as experiential evidence to support their claims. Where ascriptions are more highly contested, in other words, we can expect direct perceptual experience or feelings to play a correspondingly larger role.

Although the sui generis approach viewed religious experience as the central feature of religion in general, an ascriptive approach paints a different picture. Viewed ascriptively, claims based on experience clearly play an important role in the basic process of marking things out as special. These experiential claims, however, are not self-evident and are invariably contested. Groups that have reached high levels of consensus regarding the evidential value of the experiential claims that matter to them often elide this ambiguity.

There is, moreover, a sharp distinction between setting something apart as special and a religious or spiritual tradition. The latter are much more complicated formations that rely on composite ascriptions, such as the specification of a special path, that allow for the extension or re-creation of the original thing set apart as special in the present. In these more complicated formations, claims based on direct perceptions or immediate feelings play a more limited role. Within groups that have achieved a high degree of consensus on underlying ascriptions, subjective experience is normally subordinated to experience as defined by the group and realized formally through group practices.

Within the study of religion evidential appeals to experience have often been conflated with the embedded understanding of experience characteristic of relatively closed, self-authenticating systems. Critics of the emphasis on experience in the study of religion are right to note this distinction and in that sense to counter an overinflated view of the role of "religious experience" in the study of religion. In countering an overinflated view, however, we should not lose sight of the multitude of situations in which evidential appeals to experience are made and play a crucial role, especially in the emergence of new or dissenting movements. An ascriptive approach allows us to make needed distinctions rather than abandoning the concept altogether.

By deferring the discussion of experience per se in order to focus first on deeming things religious, we have been able to make such distinctions and paint a more nuanced understanding of the place of experience in simple and composite religious ascriptions. Having made these distinctions, let us now turn directly to the concept of experience and in so doing to consider the claim made by critics of the attributional approach that some experiences are inherently religious.

CHAPTER TWO

Experience

ACCESSING CONSCIOUS BEHAVIOR

The debate over religious experience in the past few decades has been framed in terms of the relationship between experience and representation. With the discursive turn in the humanities, many humanists turned a suspicious gaze on the concept of experience, questioning whether it was possible to speak of experience at all apart from the way it is represented in and shaped by discourse. Within religious studies, Steven Katz (1978, 1983) and Wayne Proudfoot (1985) were two of the most forceful advocates of this constructivist view, which emphasized the role of language, tradition, and culture in constituting experience. Their work, along with that of many others who participated in the general discursive turn, sharply challenged the tacitly perennialist views advanced by many of the classical figures (for example, Otto, van der Leeuw, Wach, Eliade, and Smart) discussed in the previous chapter.

In the 1990s a group of scholars led by philosopher of religion Robert Forman responded with a new perennialism that claimed, contra the constructivists, that there were certain mystical experiences that shared underlying commonalities across time periods and traditions. Referring to themselves as "psychological perennialists," they singled out the "pure consciousness event" as chief among these common underlying experiences (Forman 1990, 1998, 1999). During the 1990s, scholars tended to one extreme or the other; the key features of so-called religious or mystical experiences were *either* constituted through language, tradition, or culture *or* were in some sense universal, albeit perhaps only psychologically. For those holding to the former viewpoint, meaning was attributed to experience discursively; for those holding to the latter viewpoint, meaning was inherent in the experience itself.

Then, in 2000, in a special issue of the multidisciplinary *Journal of Consciousness Studies*, a journal he cofounded with three others in the early 1990s, Forman called for "a truce in the twenty-years' . . . war . . . between constructivists and perennialists in the study of religion." He claimed that both sides had made some good points, but he seems to have sensed that the debate had reached a dead end. In calling on scholars of religion to drop their swords, he urged them to start reading more broadly in the burgeoning research on consciousness. Fearing that scholars of religion were in danger of "painting [themselves] into a methodological corner," he encouraged them to start drawing upon the literature in neuropsychology, cognitive neuroscience, artificial intelligence, artificial life, psychology, and other disciplines to "explore how consciousness functions and how it may play a role in the constitution of reality, in spiritual experience, in the generation of doctrine, and in ritual and meditative life" (Andresen and Forman 2000, 7–8).

Acknowledging the wisdom in Forman's advice, I do not plan to address the perennialist-constructivist debate directly in this chapter, but rather to focus attention on the ways in which fields other than religious studies can cast light on the relationship between experience and representation. In turning to other fields that are studying experience, ranging from philosophy to the neurosciences, it immediately becomes obvious that discussions of experience in religious studies have been hampered by a lack of precision regarding what we mean by "experience" and a resulting inability to consider how we might access it with much rigor. Attention to these matters will allow to us set up more precisely specified objects of study and to return to the question of experience and representation with renewed insight.

We can get a glimpse of the problem simply by considering the implications of the shift from "religious experience" to "experiences deemed religious." Although both formulations maintain the adjective "religious," the noun that is modified shifts from the singular ("experience") to the plural ("experiences"). This is a significant change. Although an individual might say that they had *a* religious experience, "religious experience" without a definite article denotes something more general and abstract. Experiences deemed religious, on the other hand, unequivocally refers to *experiences* in the plural and thus to discrete experiences.

If we stop to reflect, we can distinguish between at least three different ways we use the word "experience": (1) specific experiences of something ("I experienced something" or "I had an experience in which" or "the experience was special"); (2) experience as a cumulative abstraction ("my experience suggests" or "in my experience"); and (3) types of experience, some more abstract and some more concrete ("religious experience" or "human experience" or "life experience" or "outdoor experience" or "work experience"). In addition, we sometimes use "experience" as a synonym for "consciousness" ("I was out cold [or I was sound asleep] and did not experience a thing"). This chapter focuses on experience in the sense of "experiences of something" and "experience" as a (rough) synonym for "consciousness." We will not be concerned with experience as a cumulative abstraction and will be concerned with types of experience only in so far as they can be translated into experiences of something—for example, an experience of the outdoors, an experience at work, or an experience deemed religious.

CLARIFYING THE CONCEPT

In scholarly discussions outside religious studies, the concepts of "experience" and "consciousness" are closely related and sometimes used interchangeably. Considering them together will allow us to clarify the concept of experience in light of work on experience and consciousness by philosophers and neuroscientists. Three sets of distinctions—between transitive and intransitive consciousness, first-order and higher-order consciousness, and conscious and unconscious processing—will allow us to characterize experiences with more precision.

Two Types of Consciousness: Transitive and Intransitive

In *Philosophical Foundations of Neuroscience* (2003), Maxwell Bennett and Peter Hacker distinguish between two widely recognized ways of conceptualizing consciousness—transitive and intransitive—and locate experience as a subset of transitive consciousness. "Transitive" and "intransitive" are grammatical terms that distinguish between verbs that take an object and those that do not. When applied to consciousness, they allow us to distinguish between consciousness as a state of being (intransitive) and consciousness of something (transitive). Intransitive consciousness has no object; transitive consciousness does. Intransitive consciousness is either present or absent. It is absent if we are dead, comatose, or in dreamless sleep. A person or animal loses consciousness upon falling asleep or fainting or being anesthetized and subsequently recovers it upon waking up. We can readily observe whether a person or animal is conscious in this sense. Humans and animals can feign sleep but cannot feign consciousness (244, 246–47). Upon becoming *aware of* something, including our own awareness, we become transitively conscious. Thus, Bennett and Hacker indicate: "My own (intransitive) consciousness is not an object of possible experience for me, but a *precondition* of any consciousness" (247; emphasis added).

Because experience is a vague concept, whether consciousness of any thing should count as an experience depends on how broadly we define experience. Bennett and Hacker resist equating them. They balk at the idea of describing thinking, believing, remembering, intending, and meaning as experiences, preferring to limit experience to perceiving in its various modalities and forms, sensation and bodily feelings, moods and emotions, and other things that "a person may have, enjoy or undergo, including activities and adventures" (265). Given their reluctance "to stretch the term 'experience' beyond [what they consider as] its already generous and exceedingly vague boundaries," they classify experience as a subcategory of transitive consciousness (263, 265). Bertram Malle, who will figure prominently in the next chapter, limits experience even more narrowly to those behavioral events that are unintended by the subject and not observed by others (Malle 2004, 75–76).

How much we include under the heading of experience is probably not crucial in general discussions as long as our meaning is clear in context. Nor is a precise definition of experience required in order to set up experience-related objects of study. As will be made clear in the final chapter, specifying the precise sort of experience we want to study is much more important than defining the concept of experience. When discussing the problem of accessing experience, it makes sense to construe experience as a rough equivalent to "consciousness" or "mind" as used in discussions of "theory of mind," because the question of how we know our own mind and the minds of others is roughly parallel to the question of how we access our own experience and that of others.

Clarifying the distinction between transitive and intransitive consciousness highlights one of the many difficulties that surrounded the concept of "pure consciousness" advanced by Forman and other psychological perennialists. References to "pure consciousness" and "wakeful but objectless consciousness" seem to suggest an experience of intransitive consciousness, something Bennett and Hacker argue is logically impossible. Moreover, by sharply distinguishing between "ordinary" and "mystical" consciousness as "experience modalities" with "markedly different epistemological structures" (Forman 1998), Forman sets these pure, mystical, objectless experiences so far apart from other experiences that it is difficult to imagine how, if at all, they could be studied. Logically, however, experiences of pure consciousness, as experiences of *something* (pure consciousness), are clearly determinate. If, as is claimed, they are experiences of awareness without an object, they are a kind of transitive consciousness in which one is conscious of "nothing" (no content) rather than "something" (some content). Such experiences, although unusual, still logically fall under the heading of transitive consciousness. Understanding them as such allows us to examine descriptions of such experiences more closely and locate them in relation to other unusual experiences that are also, for example, nondiscursive and unintended.

Levels of Consciousness: First-Order and Higher-Order

Considering experience as a subset of transitive consciousness (consciousness of something) allows us to clarify the distinction between primary or first-order and secondary or higher-order levels of consciousness. To get at this distinction, let us consider other types of transitive consciousness—being aware of something and paying attention to something—alongside of experiencing something. Note that in each case, we sometimes get more elaborate and describe ourselves as *consciously* aware of something or *consciously* experiencing something. Or conversely, we sometimes are aware of something even though we are not *consciously* paying attention to it. Used as an adjective to modify nouns such as "experience," "attention," or "awareness," "consciously" denotes a higher level of experience, attention, or awareness.

The difference between experiencing something and consciously experiencing something points to distinctions many consciousness researchers make between levels of consciousness. Researchers refer to these levels in various ways, for example: first-order and higher-order (Zelazo, Gao, and Todd 2007); primary and secondary (Edelman 1992); consciousness and metaconsciousness (Schooler 2002); and core and extended (Damasio 1999). Such distinctions are particularly crucial for thinking about consciousness in animals, infants, dreams, and unusual or altered states of consciousness (for overviews, see Allen and Bekoff 2007; Hobson 2007a, b; Pace-Schott and Hobson 2007; Panksepp 2007; Trevarthen and Reddy 2007).

It is widely acknowledged that nonhuman animals exhibit first-order or primary consciousness, which includes sensory awareness, attention, perceptions, memory, emotion, and action. Because animals are aware of things and attend to things, they, generally speaking, experience things. Although there may be a few exceptions, nonhuman animals are generally not *consciously* aware that they are doing these things. Conscious awareness, awareness of awareness, or meta-awareness, as it is variously called, is a key aspect of higher-order levels of consciousness. The highest level of consciousness, which is most likely limited to humans, depends upon language and the more complex mental functions associated with it. Higher levels of consciousness, however, are built upon primary or first-order consciousness and account for only a fraction of human mental activity (Allen and Bekoff 2007; Hobson 2007b, 436–37). Evidence of primary consciousness is evident in human infants at birth and research on infants and small children is providing an increasingly refined picture of the way consciousness develops (Trevarthen and Reddy 2007). Such research suggests that binary distinctions between two levels of consciousness may not be adequate to describe the complexity of experience at higher levels of consciousness and that further distinctions may be needed (Zelazo, Gao, and Todd 2007).

Dreams and automatic actions provide readily accessible examples of primary consciousness. When we are dreaming, we are aware but, lacking meta-awareness, we are not usually conscious that we are dreaming. Similarly, when we perform actions without thinking about them—which is how we perform most actions—we do them automatically, at the level of primary awareness. Automatic actions seem effortless. If we think about what we are doing, our enhanced level of awareness—our self-consciousness—often hinders our performance. Dreaming and ordinary automatic behaviors are not the only experiences in which we lack meta-awareness. There are other, more unusual experiences that fall on the boundary between sleeping and waking in which this is also the case, such as trance (spontaneous, ritual, and hypnotic), somnambulism (sleepwalking), and sleep paralysis. In all these instances where we lack meta-awareness, our experience seems to arise involuntarily, that is, apart from any conscious sense of intentionality (Hobson 2001, 85–112).

In a review of Forman's most detailed study of pure-consciousness experiences (Forman 1999), Daniel Merkur (2000) suggests, based on Forman's descriptions, that such experiences are not the result of "paring consciousness down to its minimum, [but of] . . . actively keeping certain selected thoughts out of consciousness." According to Merkur, the combination of "inner silence amid active sense perceptions" that Forman describes is "consistent with the dissociative character of self-hypnosis. . . . Rather than a mere emptiness, there is an actively structured area of exclusion and dissociation" (407). For our purposes, Merkur's specific conclusions are less important than the way that he compares the detailed descriptions that Forman provides in his longer work to other unusual experiences that have some similar features. As we will see in the next chapter, unusual experiences that we do not consciously intend provide fertile ground for religious attributions.

Levels of Mental Processing: Conscious and Unconscious

A great deal of mental activity of different types goes on below the threshold of awareness and thus outside our experience. Consciousness researchers refer to this as "unconscious processing" and describe it as taking place at a subpersonal level. Researchers do not always make clear or consistent distinctions between unconscious processing—that is, mental activity that does not and perhaps cannot surface to awareness—and primary consciousness, understood as awareness in the moment that is potentially available to conscious awareness (meta-awareness), but typically is not remembered because we do not pay attention to it (consciously attend to it). Automatic actions, for example, are discussed under both headings, giving rise to such seeming oxymorons as "unconscious volition" (Glaser and Kihlstrom 2005). Although the terminology used to describe these levels is neither as fixed nor consistent as we might like (partly due to difficulties measuring awareness [Merikle 2007]), the unconscious, which contemporary neuroscientists use to refer to mental activity below the threshold of awareness, should not be confused with the Freudian unconscious (Hassin, Uleman, and Bargh 2005).

For our purposes, the key thing to note is that a considerable amount of processing goes on below the threshold of awareness. Here again dreams provide a relatively clear-cut example. When we dream, we experience our dreams as surfacing into dream consciousness fully formed. They seem to play themselves out like movies in our minds. Dreams, which typically rework memories, process the memories into emotionally linked sequences below the threshold of dream consciousness. We react to what surfaces and in doing so affect the unfolding of the dream. In the absence of meta-awareness (that is, lacking awareness that we are dreaming), we do not do this intentionally (Cicogna and Bosinelli 2001). In lucid dreaming, the dreamer is aware that s/he is dreaming and can consciously attempt to affect the course of the dream (LaBarge 1985, 2000). As the case of dreams suggest, material may surface to (dream) consciousness already laden with affect and thus with tacit significance and meaning.

Researchers think of these levels along a continuum that runs from unconscious mental processes at the bottom to the higher-level mental processes at the top. They refer to processing that originates at the lower levels and surfaces to consciousness as bottom-up or data-driven processing and processing that begins at the higher levels as top-down or conceptually driven processing. The two processes take place at the same time, usually interact to varying degrees, and may involve complex feedback loops (Zimbardo 1992, 261–62). Dreams provide a clear illustration of how much mental processing can take place below the threshold of consciousness in the absence of any external stimulus. Dreams should alert us to the possibility that things may surface to awareness already imbued with an implicit sense of meaningfulness or import. Like experiences we do not intend, these experiences also provide fertile ground for religious attributions and we will return to them later in this chapter.

To sum up, any experience we can describe is an experience of something. We cannot talk about "pure experience" without making it an experience of something (even if the something is "nothing"). Experience is thus a vaguely defined subset of transitive consciousness. We can experience things at different levels of consciousness. Primary or first-order consciousness, which we share with other animals, lacks meta-awareness (awareness of awareness). At this level, we experience things without thinking about what we are experiencing. When we think about what we are experiencing we move to a higher level of consciousness. In addition, there is a great deal of mental activity that goes on below the level of consciousness, whether primary or secondary. Although we do not experience anything that occurs at this level until it surfaces to awareness, unconscious processing may be quite elaborate and may include the tacit ascription of a rudimentary sense of meaning and significance.

Accessing Experience

How do we gain access to experience? How as individuals do we know our own experience and that of others? How as researchers do we gain access to the experience of others? Is experience, and, by extension, its meaning, completely subjective, personal, and private as many religion scholars of the previous century claimed? Are subjects alone able to know their own experience directly and thus the only ones capable of saying what it means? Are others able to know it only indirectly? These questions lie at the heart of discussions about the relative value of first-person and third-person perspectives in the study of consciousness and reflect the long-standing divide between the humanities and the experimental sciences. If, as has just been suggested, we can plausibly refer to non-linguistic experience in nonhuman animals and prelinguistic experience in young humans, this would allow us to view the capacity for experience and the ability to represent one's own experience and that of others from both evolutionary and developmental perspectives. Viewed in this way, we can think of the ability to represent experience linguistically as layered on top of underlying non- or prelinguistic forms of experience. The layering of representations on non- and prelinguistic experience suggests that we might expect to find some continuity between the way something is experienced (non- or prelinguistically) and the way it is represented (linguistically) without having to equate them.

We can get at the underlying difference between the first- and third-person perspectives by thinking about how animals and young children without language gain awareness of what they and others are doing. As Barresi and Moore (1996) point out, they gain awareness of themselves and others in two distinct ways. They acquire knowledge of the other (third-person knowledge), through observing the other's physical movements—that is, primarily visually. We glean information about another person's emotions by observing their facial expressions and make inferences about what they are perceiving through the orientation of their head and the direction of their gaze. The availability of such information is determined by space and time; proximity matters and availability varies. Nonhuman animals and prelinguistic children gain awareness of their own actions (first-person knowledge) primarily from kinesthetic and proprioceptive feedback regarding their own body movement and position. This feedback allows them to distinguish between their own movements and the movement of objects in the world.

Researchers in many disciplines—philosophy, linguistics, psychology, and neuroscience—are attempting to understand how the first- and third-person perspectives come together in increasingly complex ways as consciousness develops across animal species and as language emerges in the process of human development (see, for example, Nagel 1986; Tomasello 1999; Tomasello et al. 2005; Hurford 2007). Viewed from the bottom up, it is evident that the meaning conveyed through linguistic representations is added to and expands on a prelinguistic foundation. Viewing experience in this way allows us to consider the question of how we gain access to experience (our own and that of others) and how experience acquires meaning from two interrelated perspectives: as awareness of experience arises in the body and through interaction with others.

Viewed developmentally, experience is better understood as embodied behavior, where embodiment is understood at multiple levels from the neural to the phenomenological and behavior is broadly construed to include linguistic and mental events as well as overt actions. Conceived in this way, there is no way (*pace* the perennialists) to *unequivocally* establish the meaning of experience apart from its expression in embodied behavior, linguistic and otherwise. Experience so conceived is typically expressed in a *range* of behaviors, some of which are public and provide data that can be queried and compared. *Contra* the strong constructivists, however, not all of these behaviors are discursive. Because embodied human behaviors incorporate cultural meanings and memories in complex ways, we cannot consider them as only biological, as the psychological perennialists seemed to argue, or only cultural, as the constructivists argued, but as a complex mixture of both biology and culture (Tomasello 1999; Tomasello et al. 2005).

If we view the linguistic articulation of experience as an outgrowth of embodied expressive behaviors that humans learn to specify linguistically in the context of an interpersonal developmental process, we can offer a more dynamic model of how we come to articulate our own experience and that of others. While acknowledging asymmetries between self and others (subjects and observers), such a model breaks down the rigid dichotomy between experience and representation and remains open to continued research on the nature of these asymmetries and the complex ways in which the intrapersonal and interpersonal are connected. Viewing experience in this way complements recent ethnographic work that encourages us to study "religion in events" at the interpersonal level (Bender 2003, 2008, unpublished) and provides a methodological point of contact between psychology and allied fields such as sociology and anthropology.

Experience and Embodied Meaning

Although there are competing viewpoints within the cognitive sciences, researchers from a variety of disciplines—including, for example, linguistics (Lakoff and Johnson 1980, 1989; Johnson 1987), anthropology (Csordas 1994), and psychology (Varela, Thompson, and Rosch 1993; Thompson and Varela 2001)—have been advocating an embodied approach to cognition for some time. This approach, which draws from the phenomenological tradition in philosophy, argues that mind, experience, and language all arise through the interaction between brain, body, and world. In a recent work, Raymond Gibbs (2006), a psychologist at the University of California, Santa Cruz, has synthesized much of the experimental research in support of this perspective, which researchers are now working to ground neurologically (cf. Gallese and Lakoff 2005). This approach presupposes that we can consider embodiment, like consciousness, at various levels. The most basic level is the neurophysiological. The cognitive unconscious forms an intermediate level, between the neurological level and the phenomenological level of consciousness, whether first- or second-order. The cognitive unconscious consists of all the mental operations that structure and make possible conscious experience (Gibbs 2006, 39–40).

Image schemata, as discussed by cognitive linguists such as Lakoff and Johnson (1980, 1989; Johnson 1987), illustrate one important kind of unconscious patterning that emerges in response to bodily movements, the manipulation of objects, and perceptual interactions. As prelinguistic patterns that couple brain, body, and world, they operate across sensory modalities, linking sensory motor activity in the world with mental representations in embodied schemas that we use to structure our experience (Gibbs 2006, 114–15). Meaning in this view is neither fixed nor infinitely variable, but *emerges through* and *is then is constrained by* schematic structures, such as CONTAINER, BALANCE, PATH, CYCLE, ATTRACTION, CENTER-PERIPHERY, LINK, and VERTICALITY, that are often metaphorically extended to more abstract forms of experience (Johnson 1987, 28–30).[1]

To illustrate how an image schema can take us from highly embodied to more abstract forms of experience, we can consider the way that two of these image schemas—CONTAINER and PATH—enabled us to think about religion in the last chapter. Our two kinds of ascriptions, simple and complex, each relied on a basic image schema. Things set part relied on a CONTAINER schema and paths to a goal relied on a PATH schema. The CONTAINER schema arises from our experience of our bodies as containers in which we put some things (food, water, air) and eliminate

---

[1] In *What Science Offers the Humanities*, Slingerland (2008, 151–218) makes use of Lakoff and Johnson's (1980) conceptual-metaphor theory, which is built on the idea of image schemata, and the related, but more elaborate, mental space and conceptual blending theories of Fauconnier and Turner (2002) as his primary bridges between the sciences and humanities. Readers interested in how these theories can help us to understand ancient texts are encouraged to consider Slingerland's extended analysis of Book 6 of *Mencius* (188–206).

others, from our movements in and out of bounded spaces, and from placing objects in other objects (Johnson 1987, 21–22). We draw on this physical experience of containment or things contained when we speak metaphorically of things that are set apart and metaphorically bounded by taboos and prohibitions against mixing, trading, or comparing them.

PATH schemas are routes for moving from one point to another, built on our own experience of movement, whether of a body part (for example, an arm in space) or our entire body, and our experience of tracking things moving around us. The PATH schema has a definite internal structure composed of a source or starting point, a goal or ending point, and a sequence of actions linking the source with the goal (Johnson 1987, 113–17). When extended metaphorically to religion, the goal is metaphorically transformed from a physical place (a destination) to a more abstract goal (a purpose). In other words, RELIGIONS AS PATHS TO A GOAL maps a PURPOSES ARE PHYSICAL GOALS metaphor on to the PATH schema. In doing so, however, the PATH schema gives form to our desire to achieve a religious goal (a desired state) and constrains the ways we can think about achieving it (Johnson 1987, 116–17). If we conceive of religions as paths to a goal, we then naturally find ourselves thinking in terms of sequences of actions (practices deemed efficacious) for moving from an original state to a desired state.

Experience and Interaction

The idea of embodied cognition is not the only potential bridge between mind and body or first-person (subjective) and third-person (objective) points of view. If image schemas highlight the way that meaning emerges through and is constrained by embodied patterns that link body, brain, and world, other researchers are arriving at somewhat similar conclusions by looking at the way meaning emerges through interaction. Starting from the analytical side of the philosophical tradition and relying on Wittgenstein's private-language argument, Bennett and Hacker (2003, 97) reject the idea that people first learn the meaning of psychological concepts (for example, pain, want, believe, intend) and then apply them directly to inner experiences to which they have private and privileged access. Instead, they argue, we learn to apply psychological terms to others and ourselves simultaneously as an extension of natural expressive behavior (Bennett and Hacker 2003, 101). Babies have a range of expressive behaviors that effectively convey feelings and sensations to those around them. In time, the child screams "ow" instead of crying, then later says "it hurts," and even later "I have pain." Simultaneously, the child learns that when another child says "ouch" it is grounds for saying that the other child feels discomfort. In this way, "the child simultaneously learns the two complementary facts of the use of 'pain,' its expressive (and later reportive) first-person use and its descriptive third-person use, and the complex manner in which the employment of the word in sentences is integrated into human natural *and* acculturated behaviour" (Bennett and Hacker 2003, 101–2, 321).

The question of how we access the experience of others parallels the question of how we know the minds of others—a central question in the philosophy of mind. Bennett and Hacker's solution to the former question is compatible with and extended by the second-person approach to the theory of mind recently advanced by Shaun Gallagher and others.[2] Drawing from the phenomenological tradition, Gallagher offers a new approach to theory of mind that locates our most basic ways of understanding others in embodied practices and second-person (I-you) interactions (Gallagher 2001, 85–86; 2005, 206–30). Knowledge of self and other, in this view, is acquired from a very young age through "imitation, intentionality detection, eye-tracking, the perception of intentional or goal-related movements, and the perception of meaning and emotion in movement and posture" (Gallagher 2001, 90). Bennett and Hacker's approach, although cast in terms of first-person/third-person asymmetries, is congruent with Gallagher's second-person approach. In both cases, we learn to apply words to our own experience and the experience of others through an interactive developmental process best described in terms of a second-person, or I-you, relationship.

Approaching the issue from different philosophical starting points, Bennett and Hacker and Gallagher agree that we convey a great deal about our experience to others in ways they can readily understand without the use of words. When someone groans in pain, trembles in fear, furrows their brow in puzzlement, or giggles with delight, we can come to an accurate understanding of what they are experiencing. They may, of course, try to hide their true feelings from us with greater or lesser success, but nervous twitches, facial grimaces, or blushes may give them away. Nor are we always fully aware of the ways in which we express our experiences in our embodied actions. We may give away more than we know. Unintended expressions conveyed through bodily movement, gesture, facial expressions, and tone of voice or other nonverbal sounds can all reveal things to others of which we are not aware. Because language and narratives of experience

---

[2] Advocates of a second-person approach to theory of mind, which is derived from the phenomenological tradition within continental philosophy, argue that the two dominant theories—the "theory theory" and the "simulation theory"—are overly mentalistic (Zahavi 2005, 179–222; Gallagher and Hutto 2008). The "theory theory" argues that we use a theory about the mind of others to gain access, and the "simulation theory" argues that we simulate the minds of others to gain access to them (see Gibbs 2006, 234–36).

are built on a nonverbal foundation that we acquire intersubjectively from a very young age, we convey a great deal about our experience and in turn understand a great deal about the experience of others without the use of words (Gallagher 2005, 115–130, 225–30).

Our ability to communicate by means of a variety of expressive behaviors does not mean, however, that we are immune from miscommunication, misunderstanding, or misrepresentation of the experience of others (and ourselves). Nor can we assume a rigid relationship between mental states and mental-state descriptors. States of mind do not have discrete boundaries and our attempts to capture them in language are always context-dependent and approximate (Malle 2005, 24–25). Nonetheless, locating the articulation of experience as an extension of embodied expressive behaviors (Bennett and Hacker, Gibbs) that develop in the context of second-person (I-you) relationships (Gallagher) offers a more intersubjective model of how we come to articulate our own experience and the experience of others. While acknowledging asymmetries between self and others, this model breaks down the rigid dichotomy between experience and behavior. At the same time it remains open to continued research on the nature of those asymmetries and the complex ways in which the intrapersonal and interpersonal are connected.

Varieties of Evidence

Building on the more nuanced understanding of the relationship between experience and behavior, we can identify the kinds of data we can gather about the experience of subjects, consider how collecting data affects the experience of subjects, and reassess the nature of the primary sources available to us.

We can categorize data about experience in relation to time, the way it is expressed, and the kinds of awareness that subjects and observers have of it. In relation to time, we can distinguish between real-time data, available while the experience is occurring; post hoc data, available after the fact; and prescriptive, or pre hoc, data, which sets out expectations for experience before it occurs. The experience in question may be described or not by the subject, and this description may be intended or unintended. (Subjects are normally aware of what they intentionally express but observers alone may be aware of what they unintentionally express.) Table 2.1 breaks down the kinds of data that can be gathered about experience.

As the chart suggests, there is a wide range of real-time data of which both subject and observers can be aware, though they may interpret it differently. There is also real-time data that may be available to the observer but not the subject (unintended expressions) and to the subject but not the observer (unexpressed data). Subjects express their experience in behavior, including speech, facial expressions, gestures, movement, and nonverbal

TABLE 2.1
Types of Data that Can Be Gathered Relative to Experience

Type of Data	Expression	Awareness	Reveals
Neurological data (real-time)	Formal record of neurological activity	Neither researcher nor subject experience the nonconscious mental events that the data records, though either may be aware of the data.	Preconditions of subject's experience
Observable data (verbal and other expressive behavior in real time)	Unintended expression	Of which observer may be aware while subject is not.	Disputed
	Intended expression	Of which subject and observer are both aware, though their views of it may differ	Experience of subject
Unobservable data (real-time)	Not expressed	Of which subject is aware, but observers are not	Experience of subject
Post hoc data, e.g. self-reports after the fact; collective report	Narrative or other formal representation	Retrospective awareness (memories of the event that may be confirmed or unconfirmed by others)	Experience of subject in retrospect

sounds. Subjects' recounting of their experience can typically be observed, though sometimes there is no one there to observe them. Sometimes they can conceal what they are experiencing from observers, though not always. Sometimes subjects experience things (for example, hearing voices, seeing visions, feeling presences) that others around them do not. They sometimes feel as if their experiences are indescribable, a feeling they may nonetheless seek to express—intentionally or not—in words, facial expressions, gestures, or postures.

Observers, of course, observe behavior, some verbal and some not, and draw conclusions from what they observe. They are sometimes able to observe behavioral evidence of feelings that subjects would like to conceal but cannot (for example, embarrassment, as signaled by blushing) or to which the subject seems oblivious (for example, restlessness, as signaled by fidgeting). Observers may observe behavior from various vantage points, some of which allow for interaction, whether face-to-face or at a distance, and some of which do not. Real-time experience is often expressed interactively with an eye toward the understanding and response of the other.

Neurological data—such as can be derived from neuroimaging, sleep-laboratory studies, and brain-lesion studies—provide real-time correlates of experience at the unconscious (subpersonal) level. Because it is unconscious, the mental events for which the data provide a record are not experienced by anyone, though, of course, researchers are (and subjects can be) aware of the data recorded. The important issue for research on experience is to what extent "decisions" about meaning are made tacitly at the unconscious level by virtue of the way that information is sorted, assembled, and patterned—that is, processed mentally before surfacing to consciousness. We touched on one way this can occur in our discussion of image schema and we will return to this issue later in the chapter in conjunction with the discussion of production of dream narratives.

Unobservable data, experience of which the subject is aware but does not express is, in a sense, "private" or "personal" experience, but thinking of it in this way can lead to confusion, since we all experience a great deal that we either do not express or express in subtle and largely unintended ways, simply because it is totally unremarkable. Moreover, confusing the private with the unremarkable may obscure the significance of claims about "private experience," since when we designate any specific experience as "private" or "personal," our reference to it has the paradoxical effect of making it—at least to some degree—public. When we refer to experiences as "private" or "personal," we are typically referring to experiences, whether expressed or hidden, that we view as significant and want to flag as "mine" as opposed to "yours." In other words, the designations "private" and "personal" carry rhetorical weight (Sharf 2000). They implicitly position private and personal experience over and against that which is public, shared, and in many cases normative. It may be that this oppositional character is what gives the idea of "private" or "personal" experience its generically "protestant" character. Confusing the assertively "private" with the unremarkable and unobserved has contributed greatly to the conceptual difficulties associated with experiences people consider special.

Post hoc data, often in the form of first-person narratives, provides retrospective accounts of the subject's experience based on the subject's memory (and sometimes the memories of others). Subjects may express (and recount and report) their experience in oral, written, or recorded form. They may claim that their descriptions do not do justice to their experiences. This is always the case to a degree, because descriptions are not the same thing as experiences.[3] Post hoc narratives are based on uncer-

---

[3] Bennett and Hacker (2003, 289) discuss this point at some length, pointing out that "[a] description is no substitute for an experience; the impression of a description differs from the impression of an experience."

tain and often unreliable memory. Memory is not photographic; people forget, misattribute, and elaborate information stored in memory so that multiple descriptions of the same experience will not be the same. Eyewitness testimony can be inaccurate (Loftus 1979). Research conducted in the wake of the child sex-abuse scandals has established that people can construct strikingly realistic, vivid recollections of events that did not, in fact, happen (Schachter 1995; Clancy 2005). Memory, in other words, is plastic rather than fixed and constructed rather than retrieved. This suggests the importance of combining people's post hoc accounts of experiences with other kinds of data when attempting to reconstruct real-time experiences. It also suggests that the comparison of multiple narratives of an experience from different points of view is an excellent way to examine how interpretations of an experience develop over time.

As historians and ethnographers are aware, post hoc data can take many forms. Thus, to use dreams again as an example, the first level of post hoc data is the memory of the dream experience, which could include not only the dreamer's memory but also an observer's memory of the dreamer's movements or vocalizations while dreaming. The second level would be the recounting of the remembered experience. Recounting typically is done selectively for an audience; repetition usually leads to reinterpretation and regularization. Cultural views of dreams will influence how they are recounted, heard, and received. There may be specialized interpreters of dreams in a given culture; they may be sought out and individuals may agree or disagree with their interpretation. Once the memory of dream experience is recounted, it is public and its meaning may be contested. Others may then appropriate these recounted dream experiences for their own purposes; they may re-narrate them in other contexts (for example, religious, scientific, or scholarly) for related or very different purposes. Finally, people may set certain kinds of dream experiences—for example, lucid dreams—apart as special and seek to cultivate them for various purposes. The multiple layers of post hoc data can thus be summarized as: experiences remembered, remembered experiences recounted, recounted experiences appropriated, and remembered experiences cultivated.

Collecting Research Data

Philosophers and experimentalists who study consciousness argue about the competing merits of first-, second-, and third-person approaches to data collection (Dennett 1991, 2003; Varela and Shear 1999; Shear 2007).[4] Given that their use of this terminology is confusing even to them

---

[4] This debate has been acrimonious, inconclusive, and rife with miscommunication. Velmans and Dennett have engaged in two extended email dialogues (see Velmans 2001, 2006) in which they have attempted to clarify the differences between their preferred methods:

(see Velmans 2001, 2006; Dennett 2003), I avoid references to grammatical persons in this context and rely instead on a few distinctions that seem necessary and clear: between subjects and observers; between those who interact, broadly understood, and those who do not; and between primary and secondary data. If, in addition, we distinguish between professional and nonprofessional observers, we can make the following observations about data collection by professional researchers:

- If observers are positioned or position themselves to interact with subjects, they can ask questions, engage in conversations, and observe and record behavior either face to face or at a distance (for example, by means of telephone, the Internet, or mailings). They can do so as professional or nonprofessional observers.
  □ Nonprofessional observers include family, friends, co-workers, acquaintances, strangers, and enemies.
  □ Professional observers include ethnographers, clinicians, teachers, and experimentalists.
- If observers and subjects interact, the observers may or may not try to affect the experience of the subjects. Professionals typically attempt to regulate their interactions with subjects in accord with the ideals of their profession.
  □ Some professionals try to change the subjects with whom they interact. Teachers typically try to improve their students' knowledge and expertise, while clinicians and therapists try to improve their patients' mental or physical health.
  □ Ethnographers and experimentalists typically try to minimize their effects on subjects, especially in situations where they anticipate that their observations or other techniques of taking measurements might affect what is being observed or measured.
- Observers can create different sorts of *primary documents* as a result of their interactions, including:
  □ Observers' accounts of the subject's behavior (recordings, reports, or other documentation of the subject's behavior)

"critical phenomenology" (Velmans) and "heterophenomenology" (Dennett). Their debate extends those that took place between Francisco Varela and Dennett prior to Varela's death in 2001 (see Varela 1995 and Dennett 1993). In these debates, Velmans and Dennett identify considerable common ground, articulate differences, claim that the other does not yet understand their position, and try to figure out how significant their perceived differences actually are. In Dennett (2007b), he conceded that "heterophenomenology [which he characterized as a third-person approach] could just as well have been called—by me—*first-person science of consciousness* or *the second-person method of gathering data*" (2007b, 252; emphasis in original). 4

☐ Observers' accounts of the interaction between subject and observer (field notes, transcripts, or recordings of conversations or other interactions)

- Researchers who do not or cannot interact with their subjects use primary documents—for example, historical records or other material traces—created by subjects and those observers who have interacted with them to create *secondary accounts* (for example, histories).

As the more interactive model of experience outlined above makes clear, the boundary between subjects and observers is not rigidly dichotomous. Nor is the boundary between lay and professional observers as sharp as scientists might hope or as nonexistent as some critical theorists would have us believe. The distinction between lay and professional observers is created by disciplinary practices—for example, ethnographic reflexivity and experimental protocols—that attempt to control the effects of the interaction between researchers and subjects in fields where such interactions are routine.

While our descriptions of experiences—our primary data—should reflect the understandings of subjects and nonprofessional observers (as relevant), this does not mean that researchers cannot interact with subjects and nonprofessional observers to expand the range or type of primary data available to them. Some experimental researchers have developed methods of working with subjects that allow them to expand their data by asking subjects to focus on various aspects of their experience (Varela 1996; Gallagher 2003; Gallagher and Sørensen 2006). Thus, for example, Gallagher (2003, 88) describes a procedure that some neurophenomenologists are using to train subjects to shift their attention from *what* they are experiencing to *how* they are experiencing it. The shift proceeds in three steps:

1. Suspending beliefs or theories about experience
2. Gaining intimacy with the domain of investigation
3. Offering descriptions and using intersubjective validations

In asking subjects to suspend their beliefs or theories about their experience in order to focus more closely on how they are experiencing, the neurophenomenologists are training their subjects to describe their experience at a more generic, process-oriented level suited to the sorts of comparative questions the researchers want to investigate.

REPRESENTATION AND EXPERIENCE REVISITED

Expanding the kinds of data available for accessing experience allows us to recast the relationship between representation and experience in terms of the distinction between real-time, post hoc, and prescriptive (pre hoc) data and then to consider the kinds of evidence available for some of the more unusual or contentious types of experience. In some instances, there is sufficient data to distinguish between real-time, post hoc, and prescriptive data. To what degree we can reconstruct real-time experience depends on the kind of real-time data available. The three types of experience to be discussed below—dreams, trance/possession, and meditation—allow us to examine the relationship between representation and experience at increasing levels of complexity. Dreams allow us to examine ways in which implicit meaning can be generated below the threshold of (dream) consciousness and provide a psycho-physiological framework for understanding a range of phenomena on the border between sleeping and waking. The relationship between trance and possession provides a vantage point for examining the relationship between the psycho-physiological and social dimensions of experience. And the interplay between meditation practices and textual traditions that valorize them allow us to examine the relations between psycho-physiological, social, and prescriptive dimensions of experience.

Dreams

There is an increasing variety of types of data available on dreams. In addition to traditional post hoc dream reports, there is a growing array of real-time neurological data from sleep-laboratory (polysomnogram), neuroimaging, and brain-lesion studies. Sleep-laboratory data also include some real-time observable data—for example, bodily movements or vocalizations. In the late 1980s, researchers introduced a portable sleep monitor (the "Nightcap") that allows for the collection of sleep data at home rather than in the laboratory (Mamelak and Hobson 1989; Stickgold et al. 1994; Alijore et al. 1995). Researchers can awaken subjects wearing the Nightcap and have them report on their immediately preceding experience. In one recent study, subjects carried pagers during the day and wore the Nightcap at night, dictating "mentation reports" when paged or awakened (Fosse et al. 2001). Such techniques generate post hoc reports of the subject's experience that are much more closely linked to real-time neurological data than dream reports generated upon waking from a night's sleep.

Lucid dreams, while unusual, provide a real-time link between physiological and subjective data. Although lucid dreamers' claims that they are aware of being aware (lucid) while dreaming were initially viewed with some skepticism, LaBerge (1985) demonstrated that lucid dreamers could communicate according to a prearranged signal from within the lucid dream. Lucid dreamers' ability to perform a prearranged action to signal to researchers that they are dreaming lucidly allowed them to mark the exact time of a subjective dream event. Under these unusual conditions, researchers are able to establish precise correlations between neurological data and real-time dream reports (LaBerge 2000, 962).

As an ordinary human experience whose content cannot be narrated at the time the experience is occurring, dreaming provides a basic, albeit not simple, context for examining the relationship between experience and representation. Although some philosophers, on the assumption that dreamers are not conscious, have questioned whether dreams can be considered experiences at all (Wittgenstein 1953; Malcolm 1959; Dennett 1976; Bennett and Hacker 2003), sleep research suggests that this view of consciousness is too simplistic. Using the techniques just discussed to narrow the gap between the alleged experience and its discursive representation, sleep researchers are able to speculate in increasingly refined and plausible ways about the relationship between dreaming and consciousness. In order to conceptualize phenomena in the gray area between dreamless sleep and full wakefulness, sleep researchers are finding it necessary to distinguish between different levels of consciousness.

To specify the way in which we are conscious in ordinary (nonlucid) dreams, Cicogna and Bosinelli (2001), for example, distinguish between three types of consciousness: awareness as the phenomenal experiences of objects and events (common to many living creatures); awareness as self-awareness (awareness of being oneself); and awareness as meta-awareness (awareness of being aware). In ordinary dreams, we are phenomenally aware and self-aware, but lack meta-awareness. Especially in the period between sleeping and waking, elements normally associated in REM (rapid eye movement) sleep, NREM (non–rapid eye movement) sleep, and wakefulness may be recombined to produce a range of unusual phenomena, such as sleep paralysis (motor inhibition after waking) and sleepwalking (motor activation while sleeping) (Hobson et al. 2000, 836–41).

Although researchers are developing more refined ways of discussing the nature of consciousness while dreaming, they are far from reaching a consensus on the relationship between the psychology of dreaming and the physiology of sleep—that is, between what is subjectively reported and what can be objectively measured (Pace-Schott et al. 2003). Competing theories reflect disagreements over unresolved issues, such as whether the raw material of dreams is randomly (Hobson 1998) or at least partially nonrandomly activated (Foulkes 1985; Cicogna and Bosinelli 2001) and the extent to which REM sleep does (Hobson 1998, 2000) or does not (Solms 2000) correlate with dreaming. Though all agree that dreaming involves an interplay between internally generated data and the processing of that data, they disagree on the weight that should be given to bottom-up or top-down processes and the extent to which physiological sleep data (for example, REM sleep) corresponds on a one-to-one basis with dream reports.

As a greater degree of consensus emerges, dream research promises to provide an increasingly sturdy psycho-physiological model that we can use to analyze the process whereby unusual experiences on the boundary between sleep and waking are deemed religious. There are two broad areas of research that seem particularly relevant in this regard. The first, which has to do with how dreams are mentally constructed and in the process take on meaning, can help us to understand how experience can take on implicit meaning before surfacing to (dream) consciousness. The second, which concerns the relationship between dreams and other experiences on the boundary between sleeping and waking, can help us to create a common conceptual framework for a range of experiences often deemed religious.

The models proposed by Foulkes (1985) and Cicogna and Bosinelli (2001) give particular attention to the process of dream construction, suggesting different levels of processing along a bottom-up/top-down continuum with continuous feedback between levels. Models of this sort have implications for understanding how mnemonic units, whether randomly or nonrandomly generated, can be assembled into dream events with structure, plot, and implicit meaning (see figure 2.1). In so far as much of this process is presumed to take place below the threshold of (dream) consciousness, the models offer a way of understanding how a dream unfolding in dream consciousness can be pregnant with meaning that the dreamer is not aware of having supplied (Flanagan 2000, 139–61; Cicogna and Bosinelli 2001, 34–36).

Hobson's activation-input-modulation (AIM) model (see figure 2.2) is also interesting for our purposes because it provides a way to map a wide range of "paradoxical and dissociated mental states, both normal and abnormal" that fall in the gray area between sleeping and waking (Hobson et al. 2000a, 832; Hobson 2001). According to this model, the dream state is determined by the interaction of three interdependent processes—level of brain activation (A), the origin of input (I), and the relative level of the aminergic and cholinergic neuromodulators (M). Using these three axes, Hobson et al. are able to locate a variety of phenomena in three-dimensional space, including sleep paralysis, lucid dreaming, hallucinations upon waking or falling asleep, and hypnotic trance (Hobson et al. 2000a, 836–41; Hobson 2001, 85–111).

Movement along the "M" axis across the sleep-wake cycle may explain the inverse relationship between thinking and hallucinating and provide a basis for reframing the concept of hallucinations more broadly and less pejoratively. In a series of linked experiments, Fosse, Stickgold, and Hobson (2001, 2004) found that directed thinking, defined "as any continued

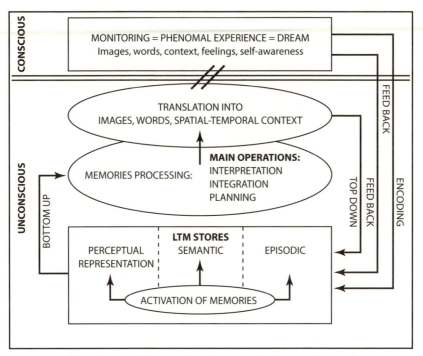

Figure 2.1. Conscious and Unconscious Processes in Dream Generation.
Adapted from Cicogna and Bosinelli (2001, 36) with permission from Elsevier.

mental effort or occupation, including contemplating, reflecting, and evaluating, as well as attempts to decide, figure out, grasp, and plan," is most common in the waking state and least common in REM (rapid eye movement) sleep. Hallucinations, defined as "internally generated (endogenous) sensation in any sensorimotor modality," though most common in REM sleep, increased throughout the night within both NREM (non–rapid eye movement) and REM sleep (2004, 299).

Traditionally, hallucinations, like bottom-up mental processing, have been defined in relation to external data. Thus, psychologists traditionally associate "real" perceptions with external data and "distorted" perceptions, such as illusions, delusions, and hallucinations, with the misinterpretation or absence of external sensory data (DSM-4, 765–68). In offering these definitions, the American Psychiatric Association notes that ordinarily the term "hallucination" is not used to refer to false perceptions that arise in dreams nor the term "delusion" used to refer to (false) beliefs widely held by a culture or religious group (DSM-4, 765–67). These exclusions, however, simply sidestep the fact that both hallucinations and

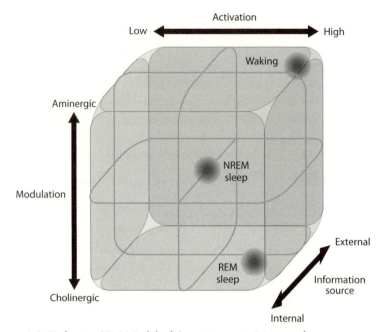

Figure 2.2. Hobson's AIM Model of Consciousness. Diagram from *Consciousness* by L. Allan Hobson. Copyright © 1999 by J. Allan Hobson. Reprinted by permission of Henry Holt and Company, LLC.

dreams arise in response to internally generated perceptions and that many religious "delusions" involve dreams and "hallucinations."

As the exclusion of dreams indicates, there is no compelling reason to assume that perceptions arising from internal sources are necessarily pathological. The pejorative (and presumptively pathological) definition of hallucinations as false perceptions artificially divides the class of phenomena that arise from internal sources and completely ignores visions that occur (remarkably frequently) in the normal population. By defining hallucinations in terms of internally generated perceptions rather than as false (that is, non-externally generated) perceptions, we can consider a wide range of phenomena including dreams and nonpathological visions together.

Possession

If the basic research on sleep and dreams offers the possibility of a psycho-physiological framework for understanding a range of phenomena on the border between sleep and waking, the relationship between trance, a commonplace boundary phenomena, and possession, a social construct, provides a good vantage point for examining the complicated relationship between the psycho-physiological and social dimensions of experience. To differentiate these aspects of the phenomena, anthropologists typically distinguish between the psycho-physiological concept of trance and the cultural concept of possession (Halperin 1996b).

With experiences that are culturally characterized as possession, we can identify a more complex range of data that can be collected than in ordinary dreaming. To illustrate, let us suppose that we have two subjects (S1 and S2), typically theorized either as a person and a spirit or as two parts of one subject (the ordinary person and a part of the person acting as spirit). As observers or participant-observers, we can collect observable data from S1 before and after possession and from S2 during possession, as well as from participants who observed the subject. These data may reflect either intended or unintended expressions of the active subject. S2 may provide real time or post hoc data regarding S1 while in possession of S1's body. S2 may, for example, offer information about the location of S1 while S2 possesses the body (real-time information) or about S1's previous behavior (post hoc). After the possession has ended, S1 can supply post hoc data about her experience—for example, regarding the extent to which she remembers what went on while she was possessed. If S1 was conscious while possessed and this was something that S1 wanted to conceal, this would most likely either remain unexpressed (unobservable data) or be expressed unintentionally.

Both ethnographic and neurophysiological researchers indicate that possession does not necessarily require trance in the neurological sense, a point that has been widely stressed in the anthropological literature (Halperin 1996b). Post hoc ethnographic interviews reveal that mediums report experiencing a range of levels of awareness while possessed and that mediums in different ritual contexts attach different values to different levels of awareness (Frigerio 1989; Halperin 1995; Buhrman 1997). Recently, a group of Japanese researchers have been able to provide neurological data that confirm these ethnographic findings. Oohashi et al. (2002) developed a portable EEG (electro-encephalographic) recording system that makes it possible to collect neurological data from moving subjects in the context of possession rituals. Preliminary tests of the EEG recording system on three participants in two Balinese ritual ceremonies identified distinct differences between the EEGs of the two participants (P1 and P2), who, based on observational data, acted and were understood by observers to be possessed during the ritual. The observed behavior of both during trance was similar, except that P2 did not stiffen or have tremors in conjunction with ritually anticipated falls during the drama, nor did he (in contrast to P1) exhibit "anterograde amnesia at any point during the drama" (436). Thus, although both were viewed as possessed, only P1 exhibited the neurological signs of being in trance.

Perhaps most crucially, we need not—indeed should not—focus simply on the possessed person but rather on the interaction between the possessed person and others. Transcriptions or recordings of interactions between "sitters" (attendees) and the "spirit controls" of entranced mediums made by late nineteenth-century psychical researchers are thus a particularly valuable primary source. Their transcripts, as they were aware, reveal the extent to which researcher-sitters were participants in an interactive process, despite their efforts to minimize their impact (Sidgwick 1915). Nonetheless, the records even of intensively studied mediums are in many ways incomplete. In the case of Mrs. Leonora Piper (1859–1950), an American medium studied by the Society for Psychical Research for more than twenty years, only a limited record was kept of what her interlocutors said to her and to one another and no effort was made to record their "involuntary exclamations, inflections, stresses, etc." Given the behavioral evidence that Mrs. Piper heard what transpired while she was in trance, even if she did not remember it afterwards, auditory cues, as psychologist G. Stanley Hall observed at the time, offered "a wide and copious margin in which suggestion can work" (Tanner 1910/1994, xix–xx). If mediums of this caliber were studied today, audiovisual recordings would allow for far more nuanced analysis of these interactions.

While trance may be considered a borderline case between waking and sleep, the neuroscience of trance—whether spontaneous, hypnotic, or ritual—is not well understood. Hobson suggests that hypnotic trance is in some ways the inverse of lucid dreaming. In hypnotic trance, we have "a dissociated state of waking into which many of the features of sleep have been inserted," while in lucid dreaming "some features of waking have been inserted into sleep" (Hobson 2001, 98–99). Moreover, the former is highly amenable to suggestion from others (hetero-suggestion) and the latter to auto-suggestion. Thus, the intention, suggestion, and cultural expectations of those with whom subjects interact shape hypnotic and ritual trance, while dreamers can suggest the direction they want their lucid dreams to take (Kahan and LaBerge 1994, 251–53).

In contrast to dreaming, where the relationship between the physiology of REM sleep and the psychology of dreaming is still a matter of debate, the psycho-physiology of trance and the social psychology of possession are overlapping but not coextensive. Although many possessed individuals are in trance, individuals can also be in trance without being possessed and possessed without being in trance. Even with the most highly hypnotizable subjects and the most gifted mediums, social interactions play a significant and sometimes determinate role in the making of the hypnotic or possessed subject (Kallio and Revonsuo 2003; Heap et al. 2004). If we view possession as a performance, it is clear not only that some performances are undergirded by trance and some are not but also that the depth of physiological trance can vary from performer to performer and performance to performance. Since the presence of physiologically verifiable trance does not determine whether a performance is successful or not, historical and ethnographic work is required to discover the cultural distinctions that subjects and observers use to decide if phenomena are authentic, fake, or pathological in various contexts (Halperin 1996a; Taves 1999; Schmidt 2000; Bender 2008).

Meditation

Meditation, in so far as it is valorized by traditions such as Buddhism, Hinduism, and Christianity, is a more contentious and complex phenomenon than spirit possession. As with spirit possession, a variety of kinds of data are available at a variety of levels of analysis—physiological, psychological, and sociological. In addition, there is an extensive textual tradition that discusses meditation. As is the case with the physiology of sleep and the psychology of dreaming, on the one hand, and trance and possession, on the other, we cannot assume that data at these different levels of analysis correspond precisely with one another. Indeed, scholars debate the extent to which textual traditions that discuss meditation reflect actual meditation practice as well as the role that meditation practice played or plays in these traditions more broadly. To get at these complexities, we can start by distinguishing three broad kinds of data: textual data produced by the traditions; real-time neurological data; and self-report data (real-time and post hoc).

Textual data. Traditions that valorize meditation have created data of various sorts, including philosophical discussions of meditation, prescriptive guides to meditation, rituals that include meditation, and, especially in the modern era, post hoc accounts of experiences that occurred during meditation. Although the Buddhist, Hindu, and Christian traditions have accumulated an extensive textual tradition that discusses religiously defined mental states related to a religiously defined path (*mārga*), scholars dispute the extent to which this historical literature was grounded in meditative or contemplative practices as they have come to be understood in the modern era (Jantzen 1995, 304–21; Sharf 1995). Beginning in the modern era, reformers within branches of the Buddhist, Hindu, and (more recently) Christian traditions promoted the practice of meditation among the laity as a means of revitalizing their traditions (Sharf 1995). Proponents of these reform movements produced a vast literature that simultaneously provides instruction and anecdotal accounts of the spiritual and other benefits of meditation. This modern emphasis has led to widespread interest in the health benefits of meditation, the emergence of a secularized practice literature, and the use of secularized forms of meditation to treat pain, stress, and depression within clinical contexts (see, for example, Kabat-Zinn 1982).

Lutz, Dunne, and Davidson (2007, 499–521) provide a highly sophisticated discussion of the ways in which neuroscientists can, and indeed should, use historical texts to help identify claims that can be investigated scientifically. They stress the need to avoid general definitions of meditation in favor of attending to the specific practices and claims about experience advanced by particular traditions and lineages (500–501). In doing so, they also advise researchers "to separate the highly detailed and verifiable aspects of traditional knowledge about meditation from the transcendental claims that form the metaphysical or theological context of that knowledge" (502).

Real-time neurological data. Research on meditation over the past several decades has produced extensive real-time neurological data using neuroelectrical (EEG, EP, and ERP) and neuroimaging (PET and fMRI) techniques. The lack of standardized designs and the diversity of meditation traditions has resulted in considerable discrepancy in the results (Cahn and Polich 2006; Lutz, Dunne, and Davidson 2007). According to Cahn and Polich (2006, 202), the most consistent findings are changes in alpha and theta waves, which "overlap significantly with early drowsing and sleep states." Differentiating meditation and early-sleep stages is one of the current challenges facing basic meditation research. Researchers are also recognizing the importance of studies that compare meditation practices of various sorts with other methods of inducing altered states of consciousness (Cahn and Polich 2006, 202). They have collected considerable data related to the effects of meditation on psychological functioning; some is derived from psychological testing and some is self-reported data. Studies of the psychological correlates of meditation practices show some consistency, primarily related to attention, absorption, and the reduction of stress and anxiety (Cahn and Polich 2006, 200–201).

Self-reported data (real-time and post hoc). Many of the researchers who advocate the use of "first-person" neurophenomenological methods in the study of consciousness—for example, Francisco Varela and Jonathan Shear (1999)—have some connection to Buddhist meditation practices and research on meditation practices more generally. Their emphasis on "first-person" and "third-person" data reflects their interest in correlating internal experience with brain activity (Cahn and Polich 2006,

182). As with research on dreaming and possession, researchers are developing new methods designed to generate data that is more reflective of real-time experience—for example, real-time button clicks by meditators to signal predetermined types of experiences to meditators and real-time bell ringing by experimenters coupled with post hoc reports of mental activity from meditators (compare Travis and Wallace 1997, 40).

Researchers of various sorts approach these data with different interests. Viewed from the perspective of neuroscientists, research on meditation, while still having considerable ground to cover, takes its place alongside research on sleep and hypnosis as one of several very interesting avenues of inquiry into the gray area between sleep, dreaming, and waking (Cahn and Polich 2006). As such, it promises in time to reveal much more than we currently know about the nature of consciousness and the ways that it can be developed by means of various kinds of practices (Lutz, Dunne, and Davidson 2007). However, just as research on trance and hypnosis is complicated by the interplay between trance and possession, so, too, research on meditation is complicated by the interplay between meditation practices and religious traditions that valorize meditation. Although practitioners are of necessity involved as subjects, they are sometimes also involved as consultants or researchers. As proponents of modern meditation-oriented movements (for example, Transcendental Meditation, Vipassana, Zen, and Centering Prayer), they promote and make use of research findings on meditation for their own purposes. This can lead to a conflation of scientific and religious claims that parallels what we observed in relation to trance and possession.

Jonathan Shear, a philosopher-practitioner affiliated with Virginia Commonwealth University and an editor of the *Journal of Consciousness Studies*, illustrates the sorts of claims that researcher-practitioners are wont to make:

> Research of various sorts on eastern meditation traditions have led to "widespread claims that it is possible to [1] refine consciousness (and underlying physiology) to display the ground, structures, and dynamics of consciousness underlying all human experience, [2] generate unusual higher states of consciousness, and [3] enhance ordinary mental and physical functioning in the process (Shear 2007, 698, citing Shear 2006).

Although there is growing evidence to suggest that meditation practices can manipulate consciousness so as to allow people to experience unusual things, it is another matter to claim that such unusual experiences actually "display the ground, structures, and dynamics of consciousness" and in doing so constitute "higher states of consciousness." While we may be able to establish neural correlates of some unusual types of experiences, these correlations tell us nothing about the meaning and significance of such experiences. Meaning and significance will still be hashed out in the old-fashioned way between *people* who make claims regarding the meaning or significance (or lack thereof) of such experiences.

This is where ethnographic and historical data regarding interpretive practices become particularly crucial. Two lines of research would seem particularly promising. The first would focus historically or ethnographically on traditions, examining the relationship between "experiences" and the practices deemed efficacious by traditions for arriving at the goals specified by the tradition. The second would focus ethnographically on contemporary discourses surrounding "experience," examining, for example, the relationship between meditation, science, and spirituality in various social contexts. Recent work by Sharf (1995, 2005) and Bender (2008) illustrates the two approaches.

In a series of articles, some already discussed, Sharf has mounted a cogent critique of the use of the category of "experience" in religious studies in general and Buddhist studies in particular. He argues, as already noted, that contemporary scholarship has "greatly exaggerated . . . the role of experience in the history of Buddhism" and traces its privileging to twentieth-century Asian reform movements that encouraged a "return" to meditation practice (Sharf 1995, 228). Furthermore, he writes:

> Even in the case of those contemporary Buddhist schools that do unambiguously exalt meditative experience, ethnographic data belies the notion that the rhetoric of meditative states functions ostensively. While some adepts may indeed experience "altered states" in the course of their training, critical analysis shows that such states do not constitute the reference points for the elaborate Buddhist discourse pertaining to the "path." Rather, such discourse turns out to function ideologically and performatively—wielded more often than not in the interests of legitimation and institutional authority (229).

In a more recent essay, he specifies that the reference point or "goal of Chan monastic practice cannot be reduced to some private 'inner transformation' or 'mystical experience.' It lies rather in the practical mastery of buddhahood—the ability to execute, day in and day out, a compelling rendition of liberated action and speech, and to pass that mastery onto one's disciples" (Sharf 2005, 266). He analyzes the "Ascending the Hall Ceremony," depicting it as a ritual intended to demonstrate the abbot's practical mastery of buddhahood and in so doing to transform him into a living buddha—that is, an enlightened one. Enlightenment, thus, is not an experience apart from the ritual performance that constitutes it.

In keeping with the discussion of religious ascriptions in the previous chapter, the "Ascending the Hall Ceremony" can be considered as a practice (in this case a ritual) to which people ascribe efficacy in relation to a special goal (in this case exemplifying buddhahood or enlightenment). Sharf's overall argument suggests that many scholars are conflating experiences that are ancillary to the path with the formal experiential goal of the path. Thus, the altered states that Sharf notes may arise in the context of Buddhist training are not the goal of the practice. Though they may or may not be considered signs of progress in the path, they are not considered particularly special in and of themselves. Buddhahood, which implies enlightenment, is the goal of the path and set apart from lesser things. Chan Buddhists traditionally ascribe efficacy to the "Ascending the Hall Ceremony" in relation to the abbot's demonstration of the practical mastery of buddhahood and by extension enlightenment. As indicated in the last chapter, the ritual presupposes the goal, defines what it means to realize it, and provides a procedure for doing so. What counts as authentic experience, in other words, is mediated by the tradition and is not inherent in the experience itself. Such a conclusion is congruent with the tradition, since, as Sharf indicates, "[t]here is, in the end, no fixed or final referent to which terms like abbot, Buddha, or enlightenment can obtain—a Buddhist truism that is repeated ad nauseam in the abbot's formal sermons" (Sharf 2005, 267).

Even if, as Sharf argues, the contemporary valorization of experience is rather ironic when considered from the vantage point of the traditional Buddhist emphasis on the mediated and contingent character of consciousness (Sharf 2005, 260), it points to a shift in the understanding of experience that can be investigated ethnographically. Although I know of no studies that have done so directly, Bender's ethnographically based sociology of religious experience among the "spiritual but not religious" denizens of Cambridge, Massachusetts, models the sort of research that could be undertaken. Using traditional ethnographic methods, Bender has been able to map the various institutional "fields" (alternative health, metaphysical religious traditions, and arts organizations) that merge to form the contemporary spiritual scene. Through ethnographic interviews, she was able to analyze the central role that narratives of "direct religious experience" play in constituting individuals as "spiritual" or "mystical" persons. Bender analyzes these narratives as a genre, concluding that they plot experience in a way that deflects attention from its formulaic character. Thus such narratives routinely depict the experience as sui generis by emphasizing the subjects' lack of cultural pre-knowledge with respect to the experience in question, the absence of social ties that might have shaped the experience, the primacy of feeling over thought, and the failure of words to capture the experience (Bender 2008, chap. 3). This narrative genre establishes the authenticity of the experience, while at the same time obscuring the conventional features of the narrative structure. A similar type of inquiry could be pursued in relation to the expectations that spiritually minded practitioners might hold with respect to scientific research on meditation.

CONCLUSION

In reexamining the relationship between representation and experience in light of the distinction between real-time, post hoc, and prescriptive data, my goal has been to see to what extent the data allows us to reconstruct real-time experiences without falling into the trap of either over- or under-valuing the reconstructive effort. Sometimes the data permit us to do a reasonable job of reconstructing real-time experience. Neurological data may in time give rise to a better understanding of how a rudimentary sense of meaningfulness—for example, a sense of portent or narrative structure—may be tacitly incorporated below the threshold of awareness. These reconstructions and models will not give us access to experience that is either free of representation or unequivocally determined in its meaning, but will allow researchers to get closer to experiences as they occur, giving them access to experiences at more levels of processing at more points in time. The reconstruction of real-time experience thus allows researchers to better understand the process whereby people make meaning. How increased understanding of the meaning-making process will feed into the process itself is a matter for observation and research. It may affect the process in some cases and in others might not.

Contrary to much academic and popular thought, the idea of private experience, though still rhetorically powerful in many contexts, is not particularly relevant to the study of experiences people consider very special, since any experience for which anyone wants to make a claim (for example, that the experience is special, private, and/or inaccessible) has to be represented publicly, if only by the sheer fact of making such a claim. Moreover, "subjective experience"—that is, the subject's account of their experience—while more conceptually useful than "private experience," is not necessarily the only primary representation of the experience in question. Others may be present and interacting with the subject and may or may not experience what the subject experiences. If not, they may have noted aspects of the subject's behavior (for example, exclamations, facial expressions, or bodily movements) of which the subject either could not be or simply was not aware. Based on their perceptions, these observers might or might not explain the subject's experience in the same way as the subject. Someone might even have recorded initial discussions of what went on. All of this primary data, however consensual or conflictual, may be considered as representing the real-time experience and considered relevant to later post hoc accounts.

The interactionist understanding of theory of mind proposed by Gallagher and others to account for how we access our own and others' experience is echoed in the "negotiated reality" and "play" models offered by Halperin for spirit possession and by Sharf for attaining buddhahood. All point to the interactive, hence also negotiated and contested, nature of public claims regarding experience, whether in "real time" or after the fact. This suggests that any method of analyzing how claims are made needs to be interactive and, at bottom, conversationally based. In the next chapter, we will turn our attention to attribution theory—that is, to theories that account for the everyday explanations that people offer for events, including experiential events. In reworking older attempts at an attributional theory of religion, I will be particularly interested in newer approaches to attribution theory that view the attribution process interactively and focus on conversation as a point of entrée.

CHAPTER THREE

Explanation

ATTRIBUTING CAUSALITY

Although the debate over religious experience was framed in terms of experience and representation, the underlying issue that made the debate so contentious had to do with causal explanations. If language, tradition, and culture constitute experience, then experiences could be explained in socio-cultural terms; if some of the more unusual experiences are cross-culturally stable, then more unusual psychological processes and brain states presumably play a causal role as well. Scholars in the humanities generally valued linguistic and cultural explanations because they emphasized cultural differences that psychological and neurological explanations tended to obscure. In arguing for the cross-cultural stability of certain types of experiences that they construed as mystical, the neo-perennialists bucked the dominant trend in the humanities. They did so, not because they were eager to embrace the naturalism or, as they would say, the "materialistic reductionism" of the sciences, but because they were sympathetic to the idea that consciousness itself might be separable from matter and able to exist independently. In this view, consciousness itself is potentially very special, perhaps so special that it exists as an absolute apart from the body, mediated by—rather than a product of—brain processes, and highly amenable to mystical or spiritual ascriptions (Kelly and Kelly 2007; Forman 2008).

Amidst these competing and conflicting perspectives, what does it mean to explain experience? Robert Forman (2008) reported on a conference held in Freiburg, Germany, in July 2008 on the topic of "Neuroscience, Consciousness, and Spirituality" that can help us to distinguish between two possible answers. The participants, who were mostly scientists, gathered to ask if "a modern day neuroscientific, functionalist or emergentist model of consciousness [can] accommodate spiritual experiences?" What, they asked, would a model of consciousness have to look like to be "both true to our modern scientific knowledge and phenomena reported by spiritual traditions?" There are two ways to approach these questions, depending on what it means for a model of consciousness to *accommodate* spiritual experiences and to *be true to* both scientific knowledge and phenomena reported by spiritual traditions. Given the conference participants' interest in questions such as whether consciousness can verifiably exist outside the brain, how the body might "transduce" consciousness, and where the domain of consciousness comes from, it is clear that for this group of researchers accommodating or being true to spiritual experiences means determining whether scientific evidence can establish the truth of what subjects claim they have experienced. We can also, however, accommodate or be true to what subjects claim to have experienced in a weaker sense if we take subjects' accounts of their experience as a description of what needs to be explained, without assuming that we need to explain it in their terms. We can distinguish, in other words, between taking the subject's description of their experience seriously in our efforts to explain it, and adopting it, defending it, or attempting to provide scientific support for it.

Wayne Proudfoot, a philosopher of religion and one of the key figures in the constructivist-perennialist debates in religious studies, addressed this issue directly in the mid-1980s by distinguishing between two forms of reduction, "descriptive" and "explanatory." When researchers seek to explain experiences that subjects view as religious or spiritual, the task of the researcher, he argued, is to avoid descriptive reductionism. Subjects' experiences, that which the researcher seeks to explain, must be described in a way that the subjects would recognize. Otherwise what researchers are explaining is something other than what the subjects claimed they experienced. While researchers need to describe experiences in terms recognizable to their subjects, they do not need to adopt their subjects' *explanations* of their experiences. Researchers can legitimately offer explanations of subjects' experiences with which subjects would not necessarily agree. This does not mean that subjects' explanations are necessarily wrong, but rather that their explanations have to be tested alongside all other possible explanations (Proudfoot 1985). Discussions of experience and explanation thus can focus on either of two questions: how and why subjects' claims *seem true to them*, regardless of whether they are true in some larger sense, or whether the subjects' claims are in fact true in some larger sense. The researchers at the Freiburg conference were interested in the latter question; this chapter is concerned with the former.

The focus of this chapter is on when and why people explain events, including experiences, that seem special to them, whether their own or others', in religious or religion-like terms. Attribution theories, as developed by social psychologists, focus on just these questions. They attempt to explain how people explain events, specifically the causal attributions or commonsense explanations that people offer for why things happen as they do (Försterling 2001, 3–4).[1] Attribution theories, in other words,

---

[1] In chapter 1, I used "ascription" to refer to the assignment of a quality or characteristic to something in order to reserve "attribution," following the social psychologists, for causal attributions (see table 1.2).

take subjects' explanations of events as the focus of their research and attempt to explain why subjects explain things the way they do. These theories presuppose a distinction between commonsense (person on the street, folk, or naïve) explanations offered in the context of everyday human behavior, including the everyday behavior of scientists, and scientific explanations subject to experimental verification. Attribution theories thus provide a way to take subjects' descriptions of their experiences and the explanations they build into their descriptions with utmost seriousness, while at the same time distinguishing between the subjects' explanations and the researchers' explanations of their explanations.

Explaining why people explain as they do in scientific terms is not the same as explaining why something happened in scientific terms. Psychologists who study processes of attribution might ask, for example, "under what circumstances individuals (subjectively) believe that their exam performance was caused by intelligence (or lack of)," but not "how intelligence influences exam performance" (Försterling 2001, 3–4). By extension, research on religious attributions might ask under what circumstances individuals (subjectively) believe that an event was caused by a supernatural agent, for example, God, but would not ask how God influences events.

The two questions about intelligence do not exactly parallel the two questions about God, however. While there are scientific ways to assess how intelligence influences exam performance, there is no way to scientifically assess how God influences an event. While both questions about intelligence can be investigated scientifically, only one of the questions about God can be. The two questions about God thus map the distinction between studying religion and doing theology, such that questions about religious attributions can be subsumed within an experimental framework, while theological questions cannot. Whether or not researchers are able to provide experimental evidence to show that consciousness is separable from the brain, a more sophisticated understanding of when and why people attribute seemingly special events to religious or religion-like causes promises to advance our practical understanding of the way experience is understood in everyday life.

ATTRIBUTION THEORY: AN OVERVIEW

Much of the initial experimental and theoretical research on attribution was done by social psychologists during the 1970s and 1980s. In the early 1980s, psychologists recognized that attribution theories provided a theoretical bridge between cognitive theory and social psychology, which led to the formation of the new subfield of social cognition (Fiske and

Taylor 1984, 1991). By the late 1980s, European social psychologists further advanced the theory by identifying different levels at which attributions are made and at which they may be analyzed, from the cognitive to societal (Hewstone 1989). In the last decade, as psychologists linked the subfield of social cognition with the neurosciences to form the subfield of social neuroscience (Cacioppo and Berntson 2005), Bertram Malle (2005, 2006) has sought to ground attribution theory in an interactive developmental process that is compatible with Gallagher's approach to theory of mind.

Attribution theory was also quickly extended to religion. In the 1970s, Proudfoot and Shaver (1975) applied attribution theories to religious experience. Drawing on Schachter's two-factor theory of emotion (Schachter and Singer 1962), they argued that "labeling and interpretation are fundamental to religious experience," such that religious experience could be understood as an experience (state of arousal) coupled with a (religious) attribution.[2] Building on this foundation, psychologists Spilka, Shaver, and Kirkpatrick (1985) extended Proudfoot and Shaver's argument beyond emotional states to formulate a general attributional theory of religion. Other developments followed. Owe Wikström (1987) joined Spilka, Shaver, and Kirkpatrick's attributional framework with Sundén's role theory of religion; Spilka and McIntosh (1997) gave attribution theory extended treatment in their discussion of theoretical approaches to the psychology of religion; and Spilka, Hood, Hunsberger, and Gorsuch (2003) used it as the organizing framework for their textbook in the psychology of religion. Although the attributional approach to religion has been more widely embraced in psychology than in religious studies, attributional theories of religion have not kept pace theoretically with later developments in psychology. Despite the hopes of its early promoters, its potential as a bridge between subdisciplines in psychology and between religious studies and psychology has not been realized.

Spilka, Shaver, and Kirkpatrick's general theory still has the potential to perform this integrative function if it is updated in light of Malle's interactionist approach and Hewstone's specification of different levels of interaction from the intrapersonal to intergroup. Before turning to this, however, we need to consider the role that attribution theory played in the debates between constructivists and neo-perennialists among scholars of religion in the 1990s, as those debates raise key issues that an updated attributional approach to religion will need to address.

---

[2] While attribution theory does not rest on this particular mini-theory, Schachter's two-factor theory of emotion stands with some subsequent modifications as one of three major psychological theories of emotions (Zimbardo 1992, 466–67) and has not been discredited as Barnard suggested (Barnard 1992).

Looking back on those debates in light of subsequent research on consciousness, it is evident that both constructivists and neo-perennialists overidentified constructivism and attribution theory, in large part because Wayne Proudfoot was centrally identified with both. In addition to co-authoring the lead article on attribution theory and religion, Proudfoot published an influential monograph on religious experience in the mid-1980s that was informed by attribution theory (Proudfoot 1985). Then, in the 1990s, Proudfoot and William Barnard were the chief protagonists in the debate over explaining experience, with Proudfoot representing the constructivist side of the debate and Barnard the neo-perennialist.[3] We can get a clearer sense of how an attributional approach to experiences that subjects deem religious needs to be refined if we carefully consider the issues that Barnard claimed the constructivist-attributional approach did not adequately explain.

In this debate, as recapped by Barnard (1997, 100–101), the constructivists made three key arguments: (1) the subject's sense that they immediately apprehend an Absolute that they *later* interpret in light of their belief systems is false, since the subject's tacit, internalized symbolic structures are operative during the experience itself and thus shape what is felt to be immediately apprehended; (2) experiences are the result of practices that produce a shift in awareness that is then labelled cognitively based on prior beliefs and attitudes; and (3) subjects are subliminally prepared (that is, "primed") by their traditions to label mysterious or anomalous experiences generated by their spiritual disciplines as "ineffable." Citing an experience of his own that did not arise in the context of religious belief or practice, Barnard (1997, 127–28) argued that constructivist explanations of experience failed to account for experiences where the subject was not engaged

---

[3] Both Barnard and Proudfoot have read this chapter and position themselves in nuanced ways in relation to the perennialist (sui generis)/constructivist debate. Barnard would accept the perennialist, or sui generis, label if it is understood as allowing a substantial role for culture and scientific comparative investigation, while at the same time resisting scientific naturalism and leaving open the question of the extent to which nonconscious processing can account for unusual experiences (personal communication, August 1, 2007). In *Exploring Unseen Worlds* (1997), Barnard aligns himself with what he describes as William James's "incomplete constructivism." In particular he highlights "James's willingness to claim that we discover the world as much as we create our experience of it [and] his theoretical openness to a preexisting, partially-formed, autonomous 'otherness' appearing within experience" (1997, 129). Proudfoot has been willing to accept the constructivist label in these debates, but feels that his critics sometimes have overstated his position. "I meant to call attention to the active cognitive component of religious experience, the part that attribution theory captures. That is constructivist in that it is an account of the way a person construes or makes sense of what is happening to her or her sense of the world. I am willing to accept the label constructivist, but not the implication that sometimes accompanied it that the experience wasn't real—either the state that was being construed or the experience that includes the attribution and construction" (personal communication, September 10, 2007).

in spiritual disciplines, did not have the relevant internalized symbolic structures, and had not been primed by a tradition. Given Proudfoot's strictures against descriptive reductionism, these are serious criticisms. While an attributional approach does not need to agree that subjects *actually had* an immediate apprehension of the Absolute, researchers do need to provide a plausible explanation of why subjects felt *as if* they did.

To do so, I will argue that we need to abandon the constructivist axiom that beliefs and attitudes are always formative of, rather than consequent to, experience in any very strong sense, in favor of a model that takes "bottom-up" or unconscious processing more seriously. Where the last chapter undercut the complete identification of experience with the subject, this chapter undercuts the identification of attribution theory with a purely constructivist—or, in psychological terms, "top-down"—account of experience. In making this argument, I am not attempting to find a middle ground between constructivism and perennialism but to reframe that debate in terms of the interaction and relative importance of top-down (culture sensitive) and bottom-up (culture insensitive) processing in relation to particular experiences.

Although Proudfoot placed more emphasis on top-down processing than I think is warranted in light of later research, his overall approach is quite compatible with what has been presented here so far. In *Religious Experience* (1985), Proudfoot elaborated an attributional approach to emotions that prefigured and roughly parallels the argument I made in the previous chapter. Drawing on Aristotle and the later writings of Wittgenstein, Proudfoot argued that emotions are not directly and intuitively known but instead are states of arousal to which subjects and observers alike ascribe emotion labels. Although Proudfoot, like Bennett and Hacker (2003) and Gallagher (2005), wanted to break down a rigid distinction between subject and observer, Proudfoot (1985, 92–93) did not seem to recognize any asymmetry between subject and observer. Moreover, where Bennett and Hacker argue that we learn the meaning of psychological terms as an extension of natural expressive behavior, Proudfoot seems to presuppose a relatively blank internal slate. Thus, though Proudfoot does not assume that emotion labels are applied arbitrarily, he views their selection, in contrast to, say, Bennett and Hacker, as largely context-dependent. In so doing, Proudfoot tilted his attribution theory further in a constructivist or cultural direction than current research suggests it needs to go.

Proudfoot's use of the conversion account of Stephen Bradley—a nineteenth-century American evangelical Protestant who experienced heart palpitations after a religious revival and attributed them to the Holy Spirit—to illustrate the attribution process heightened the constructivist slant of his theory by giving the impression that the attributional process is necessarily a conscious one. In his response to Barnard's critique,

Proudfoot acknowledged that Bradley's account, which William James used to open his chapter on conversion in the *Varieties of Religious Experience* (1902/1985, 157–60), is "a highly unusual example that includes reference to both physiological and cognitive components." In most cases, he says, "the cues that elicit arousal also provide the cognitive labels with which we understand that arousal, and we are not normally conscious of the process" (Proudfoot 1992, 794).

While Proudfoot's argument fueled the constructivist fires of the 1990s and contributed to the growing critique of the sui generis model within religious studies, few scholars of religion followed him into psychology in order to further develop the attributive model for use in religious studies. Now, as the cognitive revolution is sweeping through psychology and is even gaining a foothold in religious studies, it is time to recover and extend Proudfoot's efforts in light of more recent work in psychology.

AN ATTRIBUTIONAL THEORY OF RELIGION

In extending the argument of Proudfoot and Shaver (1975), Spilka, Shaver, and Kirkpatrick (1985) looked at meaning making in relation to the entire range of life events in order to explain when and why events are attributed to religious as opposed to nonreligious causes. Generally speaking, they theorized that the choice is "a function of the interaction of prior expectations and beliefs ('schemata') and environmental influences and cues" (161), and thus involves an interaction between top-down (schemata) and bottom-up (sensory data in the form of environmental influences and cues) processing. They refer to schemata that supply religious and nonreligious expectancies about cause-effect relationships as meaning-belief systems (MBS). They predicted that the attributor would invoke the most readily available meaning system and would turn to less-available alternatives only if the explanation arising out of the first meaning system was unsatisfactory. Availability is understood to result from the interaction of four major factors: (1) the attributor; (2) the attributor's context; (3) the event being explained; and (4) the event's context (Spilka, Shaver, and Kirkpatrick 1985, 161; for a more detailed summary of their theory, see appendix A). In order to show how their theory expanded on that of Proudfoot and Shaver (1975), they returned to the example of Stephen Bradley, indicating that in their theoretical terms, "the *event* is the heart palpitations, the *event context* includes the immediately preceding revival, the *attributor* is Bradley (who was obviously familiar with religious concepts and language), and the *attributor's context* includes Bradley's home environment and the fact that he was alone" (169).

As a basis for working through their theory in light of more recent developments, I, too, will return to Stephen Bradley's narrative, but in order to address the concerns of the neo-perennialists, I will supplement it with the narrative offered by William Barnard to illustrate his difficulties with Proudfoot's approach (for the complete texts, see appendix B). Both are post hoc accounts, written down, especially in Barnard's case, at some remove from the experience itself. We have no supporting primary data from observers present when the experiences occurred, so any attempt to reconstruct the real-time experience of either is subject to the vicissitudes of the narrators' memory and their purposes in narrating the experience. Bearing these limitations in mind, the two narratives, which stood at the heart of the controversy over the attributional approach in the 1990s, are worth revisiting in order to ensure that an updated attributional approach can encompass both types of experience and as a means of illustrating what can be gleaned from historical documents.

Events and Attributors

Let us begin by using these two cases to illustrate in more detail the four major factors (event characteristic, event contexts, attributor characteristics, and attributor contexts) that, according to Spilka, Shaver, and Kirkpatrick (1985), interact to determine which meaning-system is most available and thus most likely to be invoked by the subject to explain his or her experience. The two narratives take somewhat different forms. Bradley's narrative, which was published as a pamphlet soon after the event, has a narrative-within-a-narrative structure such that the inner narrative can be described as the event and the outer narrative as the event context. Barnard's narrative is recounted in an academic publication in order to illustrate a point many years after the event and is framed by a rationale for offering "personal testimony" in an academic context. Although this larger frame will be considered in more detail later, it provides a few extra details about the event that are incorporated here.

Event characteristics. In Bradley's case the inner narrative, which is set apart in the text with the phrase "I began to be exercised by the Holy Spirit . . . in the following manner," begins with heart palpitations ("I began to feel my heart beat very quick all of a sudden") and ends the next morning when he went out "to converse with [his] neighbors on religion" (James 1902/1985, 157–60). In Barnard's case the event was "a direct and powerful experience . . . that [he] was a surging, ecstatic, boundless state of consciousness."

Suddenly, without warning, something shifted inside. I felt lifted out-side of myself, as if I had been expanded beyond my previous sense of self. In that exhilarating and yet deeply peaceful moment, I felt as if I had been shaken awake. In a single, "timeless" gestalt, I had a direct and powerful experience that I was not just that young teenage boy but, rather, that I was a surging, ecstatic, boundless state of conscious-ness (Barnard 1997, 127–28).

Event context. For Bradley, the event context was a Methodist-led re-vival in his neighborhood that recast his earlier conversion experience as inauthentic and demanded a more direct sense of assurance by the Holy Spirit as the sign that one was truly saved. For Barnard, the event context was an "obsession with the idea of what would happen to [him] after [his] death." Throughout the day he attempted "to visualize [himself] as not existing . . . [he] kept trying, without success, to envision a simple blank nothingness."

Dispositions of the attributor (Bradley). Bradley, though previously converted at the age of fourteen, had been challenged by more recent con-verts who believed that his lack of confidence with respect to his salvation meant that his conversion was inauthentic. He attended the preaching services and conceded that he did not "enjoy religion"—that is, had not been authentically converted and thus saved the first time around. In other words, Bradley revealed not only that he had been challenged by the new meaning system, but that he accepted his interlocutors' contention that his first conversion experience was inauthentic and that he therefore did not "enjoy religion." Bradley thus was clearly disposed to experience a second conversion that would meet the (new) criteria for authenticity.

Dispositions of the attributor (Barnard). Barnard was also obsessed with the idea of what would happen to him after death. He seems to have believed that his self-awareness would not exist after death and wanted to comprehend what it would be like for that to be the case. He spent much of the day trying to visualize this and failed. Later in the day, while brooding on this problem, he had the experience in question. At the time he felt the experience was "profound" but had no other way to make sense of its "inchoate content." Nor did his "brief and, to [him], incred-ibly boring Sundays in church . . . help [him] in [his] subsequent attempts to come to grips with this mysterious and yet powerful event." It seems that Barnard's obsession was informed by a secular meaning-belief system in which it was assumed that self-awareness does not exist after death. Given his church attendance and the American context more generally, he undoubtedly had a secondary meaning-belief system, which claimed that self-awareness would survive death, at his disposal as well. This secondary belief system was most likely cast in terms of the immortality of the soul. Neither the primary nor the secondary meaning-belief system proved sufficient to explain his experience, however, and it was not "until many years later, after several years spent practicing meditative disciplines and studying Eastern philosophical scriptures, that [he] was able to give this experience a viable interpretative structure" (Barnard 1998, 171).

Context of attributor. Bradley was at home when his experience occurred. The meeting took place in the evening and Bradley went directly home afterwards and "retired to rest soon after [he] got home." The heart palpitations thus occurred in bed before he went to sleep. Although Bradley was alone in his bedroom, he was not alone in the house. His groans in response to the heart palpitations led his brother to come to his room to ask if he had a toothache and he felt the need to consult the Bible to confirm his interpretation of his experience before recounting it to his parents at breakfast.

Barnard's experience occurred while he was walking home alone from school brooding about what it would be like to die. He was alone when the experience occurred and did not discuss it with anyone ("I knew enough to remain quiet about this experience with my parents and even friends" [Barnard 1998, 171]). The puzzling nature of the experience prodded him to "find some philosophical framework that could do justice to the inchoate content" (171).

From Barnard's perspective the key difficulty with Proudfoot's contextualist-attributional theory is that he had "an experience without any real religious preparation, that possessed inherently 'mystical' qualities." It was only *after* having this [puzzling] experience . . . [that he] began to search for an intellectual framework that could accurately reflect the content that was latent in that experience." He recognizes that he was not "a completely 'lank slate" (culturally), but he rejects the idea that his "highly rudimei ary conceptual framework created [the] experience" (Barnard 1998, 171, Barnard wants us to recognize (1) that his previous theological views were not sufficient to account for the experience, and (2) that "mystical experiences are actually dynamic, ongoing, subtly shifting, complex, and often obscure processes of awareness" that may challenge our expectations and give rise to novelties that we may not be able to capture in words (170).

If we substitute "some experiences" for "mystical experiences," I think Barnard's points are well taken. Barnard's previous views could not adequately explain the novelty of his experience, which suggests that a thoroughgoing constructivist view is not adequate. Although Barnard acknowledges that he was not "a completely blank slate" culturally, he is right to insist, especially in light of the cross-cultural similarity between experiences of this type, that culture cannot adequately account for the shape of his experience. Barnard is also right to insist, following William James, that puzzling, inexplicable experiences (which he and James both view as upwellings from the unconscious) may introduce novelty and precipitate radical changes in individual lives and belief systems. Such experiences may also be linked to cultural creativity and innovation. These possibilities not only need to be conceded, they need to be more fully theorized.

There is nothing in Barnard's experience, however, that undermines an attributional approach as understood by psychologists. Barnard clearly sought to explain it. That he initially referred to his experience as "profound" and "puzzling" and only later as "mystical" reflects his search for explanations and thus situates the event within an attributional process. It took him many years of meditating and studying Eastern philosophical traditions before he characterized his experience as "mystical" and assimilated it to a meaning system that ascribes inherent features to "mystical experience" (Barnard 1997, 129; 1998, 170–71).

Although Barnard's experience—in contrast to Bradley's—exhausted the limits of his available meaning systems and precipitated a search for a more adequate explanation, this, too, can be accommodated within an attributional approach. In forging the initial theoretical link between attribution theory and meaning making, Spilka, Shaver, and Kirkpatrick (1985) emphasized the synchronic elements that inform an attribution rather than the diachronic process, in which attributions are made and modified over time. Since then, however, psychologists have elaborated on diachronic aspects of the meaning-making process (Silberman 2005), particularly in relation to the negative challenges that crises can pose to meaning systems (Park and Folkman 1997; Park 2005a, 2005b) and in relation to processes of religious conversion and spiritual transformation (Paloutzian 2005). Though psychologists have devoted more attention to the role of external (for example, the death of a loved one or natural disasters) than internal events in initiating a search for meaning, Barnard's drawn-out process of attribution nonetheless fits nicely within the general search-for-meaning framework.

In sum, in developing a general attribution theory of religion, we need to bear three points in mind. First, we can and should distinguish between an attributional approach and constructivist claims about experience. Attributions may be applied to a range of experiences from highly culturally sensitive experiences involving extensive top-down processing to relatively culturally insensitive experiences processed largely from the bottom up. People may attribute meaning to physiological sensations that are common (for example, heart palpitations) or uncommon (for example, out-of-body experiences). Barnard's experience of himself as a "boundless state of consciousness" suggests a radical disruption of his ordinary sense of embodiment, undoubtedly initiated below the threshold of consciousness. Neuroscientists have recently identified the regions of the brain that govern the sense of embodiment (Blanke et al. 2004, 2005; Arzy et al. 2006) and are now able to experimentally induce rudimentary out-of-body experiences (Ehrsson 2007; Lenggenhager et al. 2007). Subjects may characterize their experiences as novel either because they are, in fact, novel, as in Barnard's case, or because, as Bender indicates, doing so makes them seem more authentic. An attributional approach needs to be able to accommodate a range of possibilities that are not necessarily mutually exclusive.

Second, we need to distinguish between attributional analysis and attributional theory. The former analyzes how people explain events (a descriptive task), while the latter attempts to explain when and why people explain events (an explanatory task). In terms of the former, I have been arguing that we (especially in religious studies) need to be more attentive to situations in which people's ability to explain breaks down. Psychologists have been very attentive to the breakdown, reconstitution, and transformation of meaning systems in the context of traumatic life events and religious/spiritual transformation. Scholars of religion need to demonstrate this same sensitivity in relation to accounts of experiences that subjects say they cannot explain, cannot adequately express, or which seem contentless. This does not mean, of course, that we need to adopt the explanations that subjects eventually offer to explain their experiences, any more than our colleagues in psychology do. From a naturalistic theoretical standpoint, I am suggesting that constructivist theories have been insensitive to the distinction between top-down and bottom-up processing and the differential role of cultural input along the gradient that interrelates them. If many of the experiences that people consider religious or mystical emerge from the bottom up, and thus are relatively (though of course not totally) culturally insensitive, this might explain (in part) why many who have had such experiences are resistant to the constructivist account.

Third, we need to be more sensitive to experiences that are genuinely creative and generate new insights and, in some cases, entirely new meaning systems. Such experiences might fall anywhere along the processing continuum from top-down to bottom-up. Wherever they fall on that continuum, we need a deeper understanding of the ways in which seemingly novel experiences not only inspire individuals to search for alternative meaning systems, as in Barnard's case, but also lead to the generation of new meaning systems that appeal to multitudes of followers, as in the case of Siddhartha Gautama or Joseph Smith.

When supplemented by the work of Park and Paloutzian on the breakdown of meaning systems, Spilka, Shaver, and Kirkpatrick's general attribution theory of religion helps us to predict when familiar religious attributions will be applied and when a search for new meanings will be initiated, and thus to account for an important difference between the Bradley and Barnard narratives. This line of research has certain limitations, however. First, it focuses on the explanations of individual actors without paying much attention to explanations offered by observers or explanations that emerge in the context of interactions. It focuses, in other words, on cultural meanings available to actors but tends to ignore the interactive, often disputational, social context in which such meanings are created, negotiated, and deployed. Second, it is rooted in traditional social psychological research on attribution that was framed without sufficient attention to tacit conceptual distinctions, such as agency and intentionality, that inform when and why attributions are made. Malle's (2004) recent reframing of attribution theory in light of folk psychology addresses both these issues.

Although psychologists traditionally used the terms "commonsense," "folk," and "naïve" fairly casually, cognitive scientists are now using "folk physics," "folk biology," and "folk psychology" as technical designations for evolved processes of categorization that are relatively stable across cultures. These categorization processes, adapted for life in the everyday world, are pragmatic, efficient, and effortlessly acquired. They stand in contrast to scientific knowledge, which is rationalized, subject to experimental verification, and acquired with a great deal of effort. Folk psychology presupposes the ability to distinguish agents from nonagents and intentional from unintentional actions and thus a tacit understanding of the concepts of agency and intentionality (Malle 2004, 29–36).

These distinctions, Malle (2004, 60–62) argues, give rise to two kinds of explanations: "reason explanations" for events that are understood as intended and "cause explanations" for those that are not. When I do something intentionally, I can give reasons why I did what I did. Likewise, if I know someone else acted intentionally, I assume they had reasons for doing so and can speculate about them, even if I am not sure what their reasons were. In the case of unintended actions, there are no reasons to give, only causes of various sorts that can be used to explain the behavior or experience. Malle's distinctions provide a more comprehensive framework for understanding when, how, and why people choose to explain things. They are particularly helpful with respect to unusual experiences since it is often the unintendedness of the experience that leads people to make religious attributions.

Malle (2004, 239–54) has worked out an elaborate coding scheme to allow researchers to create standardized analyses of people's explanations of their behavior. In the section that follows, I use a simplified version of Malle's coding scheme to analyze Bradley and Barnard's accounts. The key features of his coding scheme are as follows:

Unintended behavior. Unintended behaviors or experiences have "cause explanations." Whether or not a behavior is unintentional is determined from the point of view of the explainer. Causes may arise from within the person (agent causes), as in the case of emotions, perceptions, personality traits, or passive behaviors such as dying; from impersonal factors (situation causes), such as the weather or physical obstacles; from the states or attributes of another person (other agent cause); or a combination of these. Agents may or may not be aware of the behavior being explained, as, for example, when an observer says, "he probably grimaced [unintended behavior] because he didn't like how it tasted [agent cause—perception]." Observable behavior, for example, facial expressions and gestures, of which the subject is often not aware (see table 2.1), provides a fertile source of such cause explanations.

Intended behavior. "Reason explanations" explain intentional actions by indicating the things the agent considered when deciding to act. Reason explanations describe the subjective mental state (for example, desires, beliefs, valuings) of the agent at the time the agent acted. Reason explanations typically employ or imply verbs that describe mental states, such as "want," "need," "fear," "hope," "think," "realize," et cetera, as, for example, "she drove to school [intentional behavior], because she didn't want [verb—desire] to be late for class [reason]." For an actor to give a reason explanation, he/she must be at least dimly aware of the reason at the time of acting and view it as reasonable grounds for acting.

Additionally, Malle's framework takes both actors and observers into account and distinguishes between situations in which actors and observers explain things to themselves (privately) and to others (interactively). The different kinds of information available to actors and observers also explains why actors are particularly likely to focus attention on explaining unintended and unobservable events (experiences), while observers are more likely to focus their attention on intended and observable ones (actions). Since unintended and unobservable events comprise much of what we typically have in mind when we refer to "subjective experience," Malle's approach explains why these types of events are particularly salient for actors and less so for observers (Malle 2004, 75–80).

Malle (2004, 63–72) also differentiates between private explanations, which are primarily concerned with making sense of things for the explainer (meaning making), and communicative explanations, which attempt to explain things to someone else and in the process manage the relationship between the explainer and explainee (behavior management).

Communicative explanations thus tend to focus on things that the explainer thinks the explainee will want or needs explained. This distinction alerts us to watch for shifts in explanatory strategies depending on whether a first-person narrative is personal (for example, a diary) or public (for example, a testimony)—that is, generated with or without an audience in mind. It would also lead us to expect that primary documents that are generated interactively may differ from documents generated by individuals for personal purposes.

Analyzing Everyday Explanations (Descriptive Analysis)

If we read Stephen Bradley's narrative as a process, the seemingly straightforward analysis of his heart palpitations as an event that he attributes to the Holy Spirit not only appears much too simplistic but also obscures his underlying attributional dilemma, which has to do with how one knows *and can demonstrate to others* that one's attributions are actually correct. His underlying dilemma is apparent when we contrast his first conversion experience with his second. The first experience, complete with a brief vision of Jesus with arms extended, feelings of great happiness, and a transformed life, seemed more than adequate until new converts questioned him about it nine years later. The central issue was whether or not he was sure that he "had religion"—that is, knew for sure he had been saved. Although his questioners "knew they had it," he could merely say that he "hoped" that he did. It was this sense of uncertainty with respect to his first experience that led him to ask them to pray for him, thinking, as he recounts, that if he hadn't "gotten religion" yet, "it was high time [he] did."

What differed between the two events that allowed him to view the second as more certain and thus more authentic than the first? Oddly enough, given evangelical theology, the difference does not seem to be in the "fruits" of the experiences. In both cases, he felt the requisite sorts of emotions (happiness, indifference to the world, and solemnity in the first instance and happiness, humility, and unworthiness in the second) and evidenced desirable attitudes and behaviors (desire for all to feel as he did and willingness to suffer for Jesus' sake in the first case and desire to tell others of his experience and to pray for their salvation in the second). Instead the difference seems to lie in the strength of the evidence for his causal attributions to unobservable agents. In other words, he seemed to think that the strength of his evidence for concluding that the Holy Spirit engaged him in the second episode was stronger than the evidence that Jesus had appeared to him in the first. Indeed, at the conclusion of his narrative Bradley indicated that he had "discharged a duty" to testify (in effect) to his certainty that "He [God] has fulfilled his promise in sending the Holy Spirit down into our hearts" and he defied "all the Deists and Atheists in the world to shake [his] faith in Christ." His defiant coda suggests that neither he nor his friends nor deists nor atheists were convinced that Jesus had appeared to him the first time around and that he had opened himself to the conversion process a second time with those attributional difficulties in mind. It was his awareness of these difficulties that, much to the delight of attribution theorists, led him to be so explicit about his attributional process.

In both conversion experiences, his attributional difficulties were centered on unintended and largely unobservable events. In the first instance, at least in retrospect, "[he] *thought* [he] saw the Saviour, by faith, in human shape, for about one second in the room, with arms extended, *appearing* to say to me, 'Come'" (emphasis added). Seeing "by faith" implied that he was aware that others would not have seen what he saw. That he *thought* he saw Jesus, who *appeared* to speak, suggests his unwillingness to claim that he *actually* saw Jesus and heard him speak. In the second instance, his attributional difficulties were linked to explaining his unusual physiological symptoms (his accelerated heart rate, the sensible stream that entered his mouth and heart, and groans sufficient to wake his brother sleeping in another room in the absence of physical pain).

If we analyze Bradley's narrative in light of Malle's distinctions between intentional behaviors, for which reasons can be given, and unintentional behaviors, for which only causal explanations can be offered, we find numerous unintentional events and a few significant intentional ones. What is most striking, however, and perhaps more typical of contested religious attributions than others, is Bradley's effort to *explain* or justify his causal explanations when narrating his second conversion experience. This sort of justification was not present in his account of his first experience and points to his awareness of the contested nature of the causal attributions he was making. In the narration of his second experience, the attribution of causation became an intentional act in its own right and one for which he then gave reasons.

The inner part of the narrative is thus framed by two intentional behaviors: asking his friends to pray that he might "get religion" and intentionally consulting the Bible before telling his parents of the night's unintended events. The first established Bradley's overall disposition (his desires) relative to the unintended events that followed, while the second provided external verification of his interpretation of the unintended events. A series of unintentional behavioral events occurred between these two intentional behaviors, including his heart beating quickly, feeling a stream entering him, and groaning aloud. He offered "cause explanations" for the unintentional events and "reason explanations" for his cause explanations. His groaning was loud enough that it woke his brother sleeping in another room, prompting his brother to offer his own "cause explanation." Telling his parents about the night's events was accompanied by a final unintended behavior, speaking in a voice that seemed not to be his own. As with the other unintended events, he explained both the event (with a cause explanation) and his explanation (with a reason explanation).

1. Intentional behavior event = asking friends to pray for him. Reason = he wanted their aid in getting religion.
2. Unintentional behavior event = trembling involuntarily at the revival, though feeling nothing. Implied cause explanation = affected by preaching but not yet saved.
3. Unintentional behavior event = feeling stupid, indifferent when he got home from the revival. Implied causal explanation = not saved.
4. Unintentional behavior event = heart beating quickly. Cause explanation = he thought he might be ill. Cause explanation rejected because [reason explanation] he felt no pain.
5. Unintentional behavior event = heart increased its beating. Cause explanation = Holy Spirit. Cause explanation accepted because [reason explanation] of the effect it had on him [it gave rise to feelings of happiness, humility, and unworthiness].
6. Unintentional behavior event = addressing the Lord. Cause explanation = he could not help it, implying that unintended feelings prompted it.
7. Unintentional behavior event = "a stream . . . came into my mouth and heart." Cause explanation = none. Stream tacitly understood as the Spirit and explicitly described as the cause of his heart palpitations.
8. Unintentional behavioral event = groaning while heart palpitating. Bradley's cause explanation = Holy Spirit groaned within him because it desired to help his infirmities [not being saved]. Rationale for cause explanation = scriptural vision in response to his question about meaning. "As if to answer it, my memory became exceedingly clear, and it appeared to me just as if the New Testament was placed open before me . . . and as light as if some candle lighted was held for me to read [Romans 8:26–27]: . . . 'The Spirit helpeth our infirmities with groanings which cannot be uttered.'" His brother's cause explanation = Bradley had a toothache.
9. Intentional behavior event = not letting his parents know of his experience until he consulted the Bible. Reason explanation = he wanted to confirm that his explanation (based on his vision of the biblical text) was correct (based on the printed Bible) before telling his parents. When he opened the Bible, "every verse seemed to almost speak and to confirm it to be truly the Word of God, and as if my feelings corresponded with the meaning of the word."

10. Unintentional behavior event = speaking under the control of the Spirit within him. Cause explanation = "[m]y speech seemed entirely under the control of the Spirit within me." Explanation of explanation [reason explanation] = "I do not mean that the words which I spoke were not my own, for they were. I thought that I was influenced similar to the Apostles on the day of Pentecost (with the exception of having power to give it to others, and doing what they did)."

In addition to explaining his explanations, two other features of his second account stand out relative to his first: the nature of his visions and the role of scripture. Visions play a key role in the attributional process in both accounts, but the content of the vision shifts from Jesus to scripture. In his first conversion account, "[he] thought [he] saw the Saviour, by faith, in human shape, for about one second in the room, with arms extended, appearing to say to me, 'Come.'" In the second, he had a vision of a scriptural text, which he then in his second major intentional act, confirmed was in the printed Bible before recounting his experience to his parents.

In both conversions, he had a vision and feelings appropriate to the situation during and after the event, but only in the second did he offer scriptural support for his attribution of the sensations and feelings to divine agency. A scriptural vision, which could be checked against the printed text, carried a level of authority within his evangelical culture that a visionary encounter with Jesus, unobserved by anyone else, did not. This suggests that the certainty with which he recounted the second experience was not just a matter of personal certainty but was a culturally informed certainty grounded in his awareness of the group recognition of the (scriptural) evidence he could marshal in support of his attribution.

The importance of scripture in providing reasons for making causal attributions to otherwise unobserved agents is also apparent in his explanation of his attribution of alterations in his speech to the Holy Spirit. He took pains to explain that he did not mean that the Holy Spirit spoke through him as a spirit might speak through a medium, but rather that he was influenced in what he said in the same way that the apostles were influenced on the day of Pentecost.

When we turn from Bradley to Barnard, we see a number of contrasts. If we take into account the larger context in which Barnard's account of his experience appears—an academic publication—it is clear that he too presented his narrative with a communicative intent. Nonetheless, for a long time it was a hidden event recalled, I am assuming, only in memories that Barnard indicates became more "highly polished" and "intellectually informed" over time. As a narrated memory of an inexplicable and, therefore, hidden event, Barnard's experience stands in marked contrast to Bradley's. Both Bradley and Barnard focused on unintended and largely unobservable events, but where Bradley struggled to convince himself and others that his attributions were correct, Barnard was unable to make any attributions at all. As he recounts it, he experienced the behavior events (obsession about death, something shifting inside) as unintentional and as inexplicable at the time they occurred.

1. Unintentional behavior event = becoming obsessed with the idea of what would happen to him after his death. Cause explanation = explainer believes that there was no apparent reason for his obsession either (we can infer) within himself or in his particular situation. "When I was thirteen years old, I was walking to school in Gainesville, Florida, and *without any apparent reason*, I became obsessed with the idea of what would happen to me after my death" (emphasis added).

2. Intentional behavior event = attempting to visualize [a practice in which he tried to imagine a counterfactual situation] himself as not existing [not existing after death = secular cultural script]. Reason = he could not comprehend that his self-awareness would not exist in some form or another after his death.

3. Intentional behavior event = continuing attempts to visualize. Reason = he kept failing in his attempts to envision a simple, blank nothingness.

4. Unintentional behavior event = suddenly, without warning, something shifted inside. Cause explanation = the explainer seems to imply that the continued brooding and attention placed on "what it would be like to die" were the occasion of something shifting inside but does not identify either as the cause.

5. Unintentional behavior event = he felt lifted outside of himself, "as if I had been expanded beyond my previous sense of self. I was a surging, ecstatic, boundless state of consciousness." Cause explanation = none. As he had no satisfying explanation ("I had no words to make sense of [it]"), he protected and preserved the experience for further reflection ("I knew enough to remain quiet").

As Malle would lead us to expect, both actors concentrate on explaining unintentional events because those are the ones they found most puzzling. In both cases, the unintentional events took place, for the most part, in private. Barnard, who was not expecting or even hoping for such an experience—indeed says he knew nothing about such experiences—at the time it occurred, attempted to account only for unintentional events. Because he valued the experience but could not explain it, he did not share it with others. Bradley, whose hoped-for experience occurred in the context of a revival, interacted with both his brother and his parents and explained both intentional and unintentional events with an eye toward an audience that he expected would be skeptical. Bradley's explanations of intentional events were thus communicative in intent. He checked the Bible before telling his parents about the night's events as a means of managing their response to his interpretation and he explained his request for prayer in anticipation of his listeners' or readers' desire for an explanation. Most significantly, he self-consciously made causal attributions that he knew would be controversial and offered reasons to support his explanations in order to convince others that his attributions were correct.

Meta-explanations of Everyday Explanations

While both Bradley and Barnard attributed their unintentional and largely unobserved events to religious causes (the Holy Spirit and a transcendent power), scientific explanations would first look for the cause of those events in unconscious natural processes. If we re-analyze both narratives from a naturalistic perspective, we can postulate a hypothetical series of steps only some of which were consciously noted by the subjects.

1. In the wake of a sermon on the terrors of hell, "[Bradley's] feelings were still unmoved" and he felt "indifferent to the things of religion" (*feelings unmoved* and *indifference to the things of religion* were cultural markers of an unsaved condition).
2. His heart began to beat very quick all of a sudden (hypothesis: his increased heartbeat was triggered by the thought of his unsaved state, primed by the immediately preceding preaching on the terrors of hell).
3. His accelerated heart beat led to thought that he might be ill, but he rejected this idea because he felt no pain (*pain* in this case serving as a cultural marker for illness).
4. As his heart continued to beat at an accelerated rate, he began to feel happy, humble, and unworthy as he had never felt before (hypothesis: the tacit rejection of a naturalistic attribution [illness] heightened the possibility of a supernatural explanation; this possibility, linked as it would be to the possibility that he was experiencing salvation, led to feelings of happiness, humility, and unworthiness—that is, feelings culturally befitting such an encounter and as "fruits of the Spirit," also cultural markers of conversion).
5. These feelings were viewed as the effect of the increased heart rate and therefore the increased heart rate was viewed as caused by (attributed to) the Holy Spirit.
6. He addressed the Lord (=interaction with unseen agent), at which point a stream (resembling air in feeling) came into his mouth and

heart (= breath is culturally associated with the Holy Spirit) and "took complete possession of [his] soul" and filled him with "the love and grace of God" (hypothesis: the feelings, understood as signs of supernatural presence, initiated a culturally appropriate response [addressing the Lord in prayer], which in turn triggered increased awareness of heightened breathing).

7. "While thus exercised, the thought arose . . . what can it mean?" (Need for explanation.) "As if to answer it, my memory became exceedingly clear" and a biblical passage (Romans 8:26–27) appeared before his eyes and he read the words, "The Spirit helpeth our infirmities with groanings which cannot be uttered" (hypothesis: his heightened state of arousal triggered a vision in a vision-prone individual [see earlier conversion account]). The visualized passage explicitly linked his groaning in response to his increased heart rate with the groaning evoked by the Holy Spirit in the passage from Romans. He attributed the feeling of a stream entering his body and filling his heart with love by extension to the Holy Spirit. The classic mark of an authentic Methodist conversion experience was the "witness of the Spirit" with your spirit that you are a child of God—that is, saved.

The process can be further schematized as an interaction between ideas (thoughts), physiological symptoms, and feelings. In each case, the process appears in italics.

1. The idea that he was unsaved triggered alarm, which was manifest physically in increased heart rate. *An idea tacitly triggered a physiological symptom.*
2. Increased heart rate led to attribution of illness, which he questioned due to the absence of pain (a cultural marker of illness). *A physiological symptom required explanation (thought).*
3. Possibility of supernatural explanation (an idea) *tacitly* triggered feelings culturally befitting an encounter with a supernatural agent. *An idea tacitly triggered appropriate feelings.*
4. The particular feelings were those culturally understood as "fruits of the Spirit." The increased heart rate was consciously attributed to the feelings and, since the feelings were understood as "fruits of the Spirit," the increased heart rate was attributed to the Holy Spirit. *Feelings* (understood as "fruits of the Spirit") *were attributed to a physiological symptom, therefore the physiological symptom was attributed to the Holy Spirit.*
5. The attribution of the physiological symptom to a supernatural agent prompted him to address the Lord (a culturally related supernatural agent). *Attribution initiated a cultural role* (penitent praying

to Lord). *Role triggered a physiological symptom* (feeling of stream entering into his mouth and heart).

6. *Vision* of biblical passage *confirmed* that *symptom* should be *attributed to the Holy Spirit and understood as sign of conversion.*

This analysis indicates that a case that has long been taken as a classic example of a simple conscious process of attribution, in which Bradley experienced an increase in his heart rate and rejected the explanation of illness in favor of the Holy Spirit, was in fact far more complex. The process was composed of conscious and tacit thoughts that triggered both physiological sensations and feelings that were explained in terms of cultural scripts. The explanation cued a cultural role that triggered a physiological response, a vision, an explanation, and a resultant thought. The narrative of the experience was intended for an audience and included explanations of the attribution of the experience to the Holy Spirit in order to make it as convincing as possible.

The concepts of roles, scripts, and cues are borrowed from research on social cognition. Research at this level presupposes an interaction between thoughts, feelings, and physiological symptoms of various sorts. When thoughts trigger feelings and physiological symptoms or thoughts and feelings influence perceptions, psychologists refer to it as "top-down" processing. When physiological processes or perceptions trigger thoughts and feelings, it is referred to as "bottom-up" processing. Because most of these interactions take place below the level of conscious self-awareness, various research techniques—for example, priming experiments and hypnotic suggestion—have been used to investigate how tacit thoughts and feelings shape perceptions and behaviors and trigger physiological symptoms.

We can also see an interweaving of cultural ideas and physical symptoms when we dissect Barnard's narrative, but the cultural aspects within the body of the narrative itself are not as pronounced as they are in Bradley's narrative and a new element—a practice of visualization—is introduced.

1. I became obsessed with the idea of what would happen to me after my death. I attempted to visualize myself as not existing (he spontaneously initiated a practice in which he tried to imagine a counterfactual situation [not existing after death] that accorded with a secular cultural script).

2. I simply could not comprehend that my self-awareness would not exist in some form or another after my death. I kept trying, without success, to envision a simple blank nothingness (the visualization exercise generated a paradox—that is, asking self to imagine self not being able to imagine).

3. Suddenly, without warning, something shifted inside. I felt lifted outside of myself, as if I had been expanded beyond my previous sense of self (hypothesis: the paradox triggered an altered state of consciousness in which self-other boundaries dissolved and perception of self-body relations were altered).
4. I was a surging, ecstatic, boundless state of consciousness (hypothesis: alteration of self-body-other relations triggered feelings of ecstasy and exhilaration).
5. The novelty and intensity of the experience triggered a need for explanation. No satisfying explanations surfaced, so the experience was protected and preserved for further reflection. Later meditation practice and the reading of spiritual texts led to his describing the experience as "mystical" and attributing it to a higher power.

The process can be further schematized as an interaction between ideas (thoughts), practices, physiological symptoms, and feelings.

1. *Thoughts led to a spontaneous visualization practice.*
2. *Practice generated a mental paradox.*
3. *Paradox resolved itself in the dissolution of self-other boundaries.*
4. *Dissolution of self-other boundaries triggered feelings of ecstasy and exhilaration*
5. *The novelty and intensity of the experience required explanation.*

The process of attribution in this case is very different from that in the previous example. Here there is no elaborate interweaving of thought, physiological symptoms, and feelings within a highly elaborated cultural matrix of expectations. This experience is rather focused and compact. It is precipitated by the unsuccessful attempt to visualize a widespread secular cultural script (the idea that the soul/self is extinguished with the death of the body). The idea of trying to visualize the self not existing after death apparently emerged spontaneously. I am hypothesizing that the mental paradox involved in the visualization triggered the dissolution of self-other boundaries, that the dissolution of self-other boundaries triggered feelings of ecstasy and exhilaration, and that the novelty, intensity, and suddenness of this experience triggered the need for explanation.

Compared to research on roles, scripts, and cues, relatively little research has been done on the role of practices (visualization, meditation, chanting, fasting, et cetera) in triggering unusual experiences, though there is considerable historical and anecdotal evidence to suggest that this is often the case. Psychologists and anthropologists have focused attention on the effects of cultivating mental imagery (Noll 1985; Luhrmann 2004, 2005, 2007) and self-injurious behaviors, such as fasting, sleep deprivation, and flagellation (Kroll and Bachrach 2005). As noted above, neuroscientists have recently identified the regions of the brain that govern the sense of embodiment (Blanke et al. 2004, 2005; Arzy et al. 2006) and are now able to experimentally induce rudimentary out-of-body experiences (Ehrsson 2007; Lenggenhager et al. 2007), though there are as yet no studies that link practices with the manipulation of those brain areas.

FOUR LEVELS OF ANALYSIS AND ATTRIBUTION

Although an attributional model rooted in psychology has much to offer to researchers with respect to experiences subjects consider religious, it needs to be elaborated at multiple levels of analysis to intersect more fully with the various kinds of questions that interest historians and ethnographers, as well as experimentalists (Antaki 1988). With their awareness of the complex ways in which such experiences have been mobilized in the world, historians and ethnographers can challenge psychology not only to further integration across psychological subfields but to integration with disciplines that deal more fully with historical dynamics, group processes, and textual traditions

Events and attributors can be analyzed at four different levels: intrapersonal, interpersonal, intragroup, and intergroup (adapted from Hewstone 1989), where intrapersonal would include the personal and subpersonal levels discussed in the last chapter. I do not include a cultural level, since culture—if understood as (behavioral) representations of thoughts, feelings, and sensations—is present in some form at all four levels.[4] The controversies noted in the last chapter over the relationship between the personal and subpersonal levels extend to the other levels as well.

As indicated in the introduction, there are a variety of theories regarding how these levels might relate, including theories that assume that it will be possible in time to describe higher levels in the terms operative at lower levels (reductionism), theories that argue for the emergence of distinct properties at higher levels that are not present at lower levels

---

[4] Some cultural theorists argue that culture is sui generis and can be explained only in terms of culture (see Barlow, Cosmides, and Tooby 1992, 22, for examples). Others, such as Geertz (1973), contrast the cultural, which he associates with the symbolic, with the social and psychological, thus constituting symbols "as meaning-carrying objects external to social conditions and states of the self ('social and psychological reality')" (Asad 1993, 32). Here I am assuming that culture is intrinsic to signifying and organizing practices at all levels from the intrapersonal to the intergroup (for researchers that approach culture in this way, see, for example, Sperber 1996, Deacon 1997, and Richerson and Boyd 2001). Deacon (1997, 22) defines culture in terms of symbolic representation. Richerson and Boyd (2001) define culture as "information capable of affecting individuals' phenotypes which they acquire from other conspecifics by teaching or imitation."

(emergentism), and theories that argue for reciprocal constraints between levels (for an overview, see Clayton 2004, Clayton and Davies 2006). These discussions go to the heart of the mind-body problem, the relationship between consciousness and behavior, and the relationship between single and multiagent systems and certainly will not be resolved here. Although I find Jaegwon Kim's (2006) discussion of the philosophical difficulties involved in arguments for top-down causation illuminating and Murphy's (2006) suggestion for resolving them intuitively plausible, the aim of this section is to use Bradley's and Barnard's narratives to illustrate the kinds of research and analysis that could be done at each level, while leaving open the question of how the levels relate.

At each level we can distinguish between the ways subjects and observers explain an event and the ways researchers might explain it within a scientific framework (table 3.1). At each level we can also identify a central question that comes to the fore: at the interpersonal level, the relationship between subpersonal (nonconscious) and personal (conscious) processes in the attribution of meaning; at the interpersonal level, the actual or anticipated impact of intimate others on the interpretive process; at the intragroup level, the role of institutionalized decision-making processes on interpretation; and at the intergroup level, the role of processes of in-group/out-group definition and management on interpretation.

Where all four levels are reflected in Bradley's narrative, which was explicitly framed as "testimony," Barnard's narrative qua narrative was private and thus limited to the interpersonal. With its publication Barnard's entered into the public realm as "personal testimony" and most specifically into academic (that is, intragroup) controversies over mystical experience.

The Intrapersonal Level

As post hoc accounts without supporting real-time data, both the Bradley and Barnard narratives allow for only very tentative reconstruction of their real-time experiences. Nonetheless, a great deal can be learned from the tentative reconstructions generated by a close reading of such narratives, as the last section was designed to show. In particular, we can identify the kinds of experiences that triggered the need for explanation and a range of conditions in which such experiences occur. This kind of analysis can help us refine questions for further research with respect to the interplay between subpersonal (nonconscious) and personal (conscious) processes in the attribution of meaning.

Although Bradley's and Barnard's experiences both involved the attribution of meaning to neurophysiological processes (heart palpitations and unusual breathing [Bradley] and alterations in perceptions

TABLE 3.1
Explanations at Different Levels of Analysis

Level of Analysis	What Explained: Event/Experience	How Explained by Attributor	How Experience might be Explained by Researchers
Intrapersonal	• Bradley experienced increased heart rate and sensation of a stream entering his mouth and heart. • Barnard experienced an ecstatic, boundless state after trying to visualize himself not existing.	• Bradley attributed palpitations to stream and stream to Spirit. • Barnard was puzzled by his experience, ascribed significance to it, but initially had no explanation. He later understood it as "mystical."	• *Top-down processing* in which thoughts (*schemas*) informed by *cultural scripts* unconsciously triggered physiological states, feelings, and *cued* culturally defined *roles* (Fiske and Taylor). • The attempt to visualize a cultural belief (*practice*) created a mental paradox that triggered the dissolution of self/other boundaries (*bottom-up processing*), which in turn triggered feelings of ecstasy and exhilaration (*experience*). Novelty triggered the need to explain.
Interpersonal	• Bradley engaged with brother at night and parents in the morning. • Barnard did not share experience at time.	• Bradley's brother attributed it to a toothache; parents' response not given. • Barnard protected exp. from attributions of parents and friends.	Processes of attribution often take place inter-individually that is, in conversation (Malle). Individuals may solicit attributions or they may be offered unsolicited. Subjects' attributions may be accepted, rejected, or modified (Stark).
Intragroup	• Methodist interpretations. • Academic interpretations.	• Distinctions b/w old-time and modern exp. • Distinction b/w private and public discourse.	Processes of attribution take place in accordance with group norms as institutionalized in organizational leadership roles and decision-making processes (Hutchins).
Intergroup	• Methodists, other Protestants, Deists, and Atheists. • Academics and religious traditions.	• Methodists heighten or deemphasize differences with other Protestants. • Academics heighten or deemphasize differences with religious traditions.	Processes of attribution linked to the categorization and creation of in-groups and out-groups (evolutionary psychology). Management of multicultural identities via integration, compartmentalization, alternation, or synergy (social psychology of culture).

of self-body-other relations [Barnard]), Bradley's experience unfolded within an elaborate matrix of cultural expectations and personal desires, while Barnard's was relatively less encumbered. While acknowledging that claims of novelty are culturally associated with authenticity within certain contemporary movements, this should not preclude the attempt to understand how experiences might arise that are genuinely beyond the grasp of the experiencer. Thus Bradley's experience involved extensive top-down processing in which suggestion and desire—that is, thoughts (schemas) informed by cultural scripts—undoubtedly played a significant role in unconsciously triggering physiological states, feelings, and cued culturally defined roles. As such, his experience can be readily easily understood at the intrapersonal level in light of mainstream research in social cognition (Fiske and Taylor 1991).

Although Barnard's conscious obsession with death and his attempt to visualize the absence of self-awareness may have triggered the alterations in self-body-other relations, Barnard's experience involved much less overt top-down processing than did Bradley's. Though recent research on the sense of embodiment and out-of-body experiences is potentially relevant for explaining Barnard's experience at this level, we still know little about how various practices might trigger such experiences or how meaning is ascribed to such processes at the subpersonal and personal levels. Research on meditation practices may in time shed light on the first issue and research on how the sense of meaningfulness emerges in the process of dream construction may in time shed light on the second.

The Interpersonal Level

In focusing on experiences as behavioral events in which subjects and observers have access to overlapping but not coextensive information, we have extended the concept of experience beyond the individual to include those immediately present at the time of the experience. This approach, while recognizing asymmetries between subjects and observers, is designed to be open to continued research on the nature of those asymmetries and to the complex ways in which the intrapersonal and interpersonal are connected. Although constructed as narratives of individual experiences, both the Bradley and Barnard narratives illustrate the crucial role of actual (Bradley) or anticipated (Barnard) interactions with intimates at the interpersonal level. Thus in Bradley's case, although the attributional process began while he was alone in his bed it unfolded through interactions with his family members and scripture. His brother offered an alternative attribution (toothache) and, anticipating a negative reaction from his parents, Bradley armed himself with a scriptural defense before engaging them. The novelty of Barnard's experience, the fact that he had "no words to make sense of [it]," coupled—I infer—with a fear that others might dismiss it, led him to refrain from telling his parents and friends ("I knew enough to remain quiet . . ."). Later, with a meaning framework in place, he did choose to share the experience, even to the point of publishing it. Rodney Stark has suggested that the response of intimates is a crucial factor in maintaining and developing an interpretation of an experience. Indeed, he suggests that with the support of an intimate circle of family and friends novel experiences can withstand the pressure of established groups and give rise to new religious movements (Stark 1999, 296–303).

The Intragroup Level

At the group level, decisions about what is "deemed religious" are made differently than at the intrapersonal or interpersonal levels and in many cases require different methods of analysis (for example, historical, ethnographic, computer modeling). Individuals and intimate others, such as Bradley and his family members, typically make such decisions at the interpersonal level, while authorized leaders typically make them in accord with organizational norms and procedures at the group level. Social organizations, in other words, distribute the power to define situations as real and thus to deem experiences as religious or authentic. Social organizations do this in many different ways, as evidenced within Christianity, for example, by a wide range of ecclesiastical authority structures ranging from the elaborate hierarchy of the Roman Catholic Church to the minimal structures of small-scale Pentecostal groups.

That Bradley crafted his narrative with an eye to the group level is evident both in terms of its form and its content. His narrative took the form of a testimony designed to convince others that he had "gotten religion" and to convince those who had not to do likewise. The narrative is suffused with idioms that reflect an evangelical Protestant subculture and his descriptions of the revival and the recent converts' doubts regarding the authenticity of his earlier conversion experience reflect the intragroup tensions within evangelicalism generally and Methodism more particularly. Many, if not most, evangelical Protestants viewed visionary experiences in which people appeared and spoke skeptically, even, as in Bradley's case, when the person was Jesus. Visionary experiences of authentic scriptural passages that confirmed evangelical teachings were much more likely to meet with favor. Though evidence of the "witness of the Spirit," along the lines Bradley provided in his narrative, was the classical marker of an authentic Methodist conversion experience, by the time he published his account some Methodists were beginning to question the authenticity of more florid experiences of this sort and they were increasingly associated with "old-time" or "old-fashioned Methodism" (Taves 1999, 76–117). There was, in other words, intragroup dissent during this period that would lead to the marginalization of experiences of the sort recounted by Bradley over time and the formation of new groups later in the century (such as the holiness and Pentecostal movements) that would insist on the importance of such experiences.

In narrating his experience within an academic publication intended for scholars of religion, Barnard also anticipated responses he might receive at the group level. He prefaced his personal account by acknowledging his awareness that "personal testimony is often frowned on in academic philosophical discussions" and defended "this transgression of academic orthopraxy" on the grounds "that attempts to 'ghettoize' the religious experiences of scholars . . . [are] overly restrictive and methodologically unsound" (Barnard 1997, 126). Barnard's aim, however, was not to undercut the taboo against "religious testimony" for religious reasons but in service of scholarship—specifically, to negotiate a middle way between two intragroup factions within religious studies. Barnard brought his "private life" into his public academic work as evidence to support "an incomplete constructivist" position—which he described as a Jamesian middle ground—in the debate between the "perennialists" and the "constructivists" (126). Based on his reconstruction of James's philosophy of mysticism and his readings in Eastern religions, he labeled his experience "mystical" and used it to illustrate the way in which a "preexisting, amorphous, but vividly felt spiritual reality" can, in interaction with previous religious beliefs, "account for the ability of mystical experiences to bring something new into existence" (129). In doing so, he rejected an attributional approach in so far as it presupposes a strong constructivist understanding of experience. While he would not describe himself as embracing a traditional sui generis understanding of mystical experience in so far as it assumes that such experiences share a common core, *Exploring Unseen Worlds* (1997) can be read as one of a number of attempts to defend a more nuanced understanding of mystical experience.

While humanists attend to the historical processes of interpretation within groups, social scientists are using computer modeling to predict the relationship between authority structures and the degree of interpretive diversity maintained within groups (Hutchins 1995, 243–62). Researchers are linking cognitive modeling (in cognitive science) and social simulations (in sociology) in order to study collective or distributed processes of cognition and to clarify the nature of the relationship between the individual and collective levels of cognition within multiagent systems (Sun 2006). In so far as what is deemed appropriate by a group reflects a kind of organizational knowledge that transcends the individuals who produced the knowledge, the attributional process may involve emergent properties of groups that cannot be reduced to the individual processes from which they emerge (Panzarasa and Jennings 2006).

The Intergroup Level

Sociologists and social psychologists analyze the diversity of interpretations tolerated within groups and the role of categorization in the formation of in-groups and out-groups (Tajfel 1981; Brown 2000a, b; Hogg 2001; Brewer 2003). Whether leaders want to strengthen or downplay the distinctions between their group and other competing groups can affect what experiences are deemed authentic. Thus, while there were certainly internal factors that contributed to intragroup conflict over visionary experiences among Methodists, much of the discomfort that they felt can be chalked up to intergroup challenges and perceptions and their desire for legitimacy in the eyes of these other groups. Deists and atheists, as Bradley indicates, were skeptical with respect to Methodist claims about religious experience, particularly those involving visionary and bodily experiences. The more established Protestant groups had serious doubts as well, viewing the rough-hewn early American Methodists and their uneducated preachers with disdain and disparaging their claims of religious experience as mere "enthusiasm."[5] As Methodists grew in number and affluence, they sought to overcome their out-group status by requiring more education of their clergy, downplaying the experiential aspects of their tradition, and requiring more elaborate justification for experiential claims. Conversely, religious bodies that want to strengthen their group boundaries may place added emphasis on distinctive features of the group, including distinctive claims with respect to experience.

Although Barnard does not say much about his involvement with Eastern religions, apart from indications that he eventually practiced meditation and read Eastern religious texts, nor about how these activities related to his development as a scholar of religion, scholars of religion typically manage the boundary between their professional identity and their religious group affiliations, if any, with some care. As Barnard observes, "personal testimony" to one's religious beliefs is widely viewed as a violation

---

[5] Methodists (and Protestants more generally) did not refer to themselves as "enthusiasts" or "holy rollers," both of which carried connotations of false religion and were reserved for breakaway groups deemed inauthentic. Distinctions between "old-time" or "old-fashioned" and more up-to-date forms of Methodism reflect distinctions within the group, but did not connote falseness or inauthenticity. Sharf (1995) provides a parallel analysis of the controversies surrounding *vipassana* and *kensho* within Buddhism.

of intragroup norms of the academy. Religious studies scholars do not always abide by these rules and scholars who reveal their religious experience or commitments are often viewed with some suspicion or discomfort by those who do not. This intragroup pressure to relegate personal religious testimony, and by extension religious commitments and affiliations, to the private sphere tacitly defines and constitutes the academic world as secular (Marsden and Longfield 1992).

CONCLUSION

This book, like Barnard's, is positioned in relation to discussions and debates within religious studies and the academy more generally, though welcoming rather than resisting naturalistic explanations of experiences such as Barnard's. Though Barnard sees value in resisting such explanations, the disagreements between us should not be exaggerated. In response to an earlier draft of this chapter, Barnard indicated that he had no difficulty with attribution theory: "[i]f all that [it] is doing is saying that I sought to explain my experience, then fine (even if that conclusion is, to me, rather prosaic and uninteresting)." He maintains, however, that "there were qualities inherent in the experience at the time that matched [his] later understandings [of the experience as mystical]." While this understanding gave him "a language that 'fit' the experience" and "a larger framework of meaning" in which to place it, it did not in his view "change the shape or form of the experience" and, in that sense, "constitute it as mystical" (personal communication).

Nonetheless, I would stress the following differences: (1) While the framework of meaning that is applied to an experience does not necessarily change the experience in the strong sense demanded by the more extreme constructivists, the *meaning* of the experience (and in that sense, the experience as memory) does change depending on whether it is placed within a meaning frame that is secular, naturalistic, generically mystical, or tradition-based (Taves 1999, 353–58). (2) The naming of an experience as generically mystical positions the experience within a meaning framework associated with a particular reading of William James and a particular understanding of religious studies that has been criticized by some (Jantzen 1995, 305–21) and embraced by others (Barnard 1997; Kripal 2001, 2006). (3) Limiting our comparisons to experiences researchers or subjects consider religious tacitly protects them from naturalistic explanations. Those interested in explaining experiences such as Barnard's scientifically will need to compare them with other experiences in which the boundaries between self, body, and other are radically altered and that subjects do not necessarily view as religious. In so far as

Barnard and others within religious studies resist such comparisons, they inhibit the kind of integration across disciplines advocated here. In the next chapter, we will take up the matter of comparison and the distinctive possibilities associated with comparisons between things that are not necessarily considered religious and things that often are.

CHAPTER FOUR

Comparison

CONSTRUCTING AN OBJECT OF STUDY

In our analysis in the previous chapter of the way in which Bradley and Barnard attributed their experiences to the Holy Spirit and a transcendent power we distinguished between descriptive analysis, which analyzed how Bradley and Barnard explained their experience in their own terms, and metaexplanatory analysis, which analyzed the narratives from a naturalistic perspective. The latter generated alternative naturalistic explanations—flagged as hypotheses—that can be formulated as research questions. So, for example, if we take the hypothesis that Barnard's attempt to visualize the absence of self-awareness triggered an alteration in his sense of self in relation to his body and the world, we can translate it into a research question by asking if visualization practices can trigger such alterations and, if so, how are they are processed and formed at unconscious levels before surfacing to awareness.

To get at this question we would need to specify precisely the features of the experiences that interest us. Barnard's experience taken on its own is of limited help in this regard. Not only was his description very vague and stated many years after the fact, but as a single experience it does not provide any way to tease out what aspects of it, if any, might be cross-culturally stable and what aspects might have been unique to him. To establish this we would need to conduct comparative research. Ideally, we would have real-time data (or as close to real-time as possible) from people in a variety of cultural contexts and traditions. Comparison of these experiences would allow us to specify common features more precisely and then, on that basis, to explore the ways in which cultural differences emerge.

We could also ask other questions. In light of the distinction we have made between simple and composite ascriptions, we could ask, for example, if there are traditions that consider visualization practices efficacious relative to a goal they deem religious. If so, do the practices always work? If not, how do practitioners deal with or account for that? Assuming for the moment that visualization practices can trigger such alterations but do not do so reliably, we can ask whether the efficacy of the practices is linked to factors such as personality, training, or beliefs about whether the practices work. We could design research to test such possibilities.

We could also examine the relationship between experiences that individuals consider religious and traditions that rely upon composite ascriptions to make authoritative judgments about experience at the group level. What happens if someone has a seemingly spontaneously experience, which they consider significant, within a tradition or cultural context that places little value on such experiences? Will it still seem important? If it does, what claims will the individual make about the experience in light of what is or is not expected by the group? How will the group respond? Alternatively, what happens if the tradition or culture does value such experiences? Will they develop, shape, and/or constrain it? And what happens if outsiders or internal critics attribute experiences to different causes than insiders do? How, for example, do insiders respond when researchers or reformers attribute experiences they consider religious to nonreligious causes? Again, we can set up studies to explore these sorts of situations.

All these questions presuppose comparisons, some obvious and some subtle. This chapter is devoted to setting up comparisons. Comparisons are powerful research tools, but in order to make effective use of them we need to have a clear sense of the kinds of comparisons we can set up and the different sorts of questions each kind of comparison allows us examine. Here we will consider three different types of comparisons: (1) between simple ascriptions (experiences deemed religious), (2) between composite ascriptions (experience in the context of paths deemed religious), and (3) between simple and composite ascriptions. Examples of each type will illustrate the different kinds of questions that the particular type of comparison allows us to explore.

COMPARING EXPERIENCES

Historians, ethnographers, and experimentalists all do comparative work. Historical methods are particularly suited to making comparisons over time, ethnographic methods to making comparisons across cultures, and experimental methods to making comparisons between experimental and control groups. Although their distinctive methods allow them to generate different kinds of data, each can contribute to research projects set up in the three ways discussed here. In fact, as will become clear, some projects require collaboration between researchers trained in historical, ethnographic, and experimental methods. Before discussing these three ways of setting up comparative studies of experience, we need to understand some basic challenges that have been raised with respect to comparison in the study of religion.

Although the academic study of religion has been comparative from the outset, scholars have sharply criticized the traditional approaches to comparison in recent decades. Their main criticisms, as summarized by William Paden (2005), are as follows:

- Comparison suppresses cultural differences, either by imposing a false homogeneity on all its examples or by suppressing the subjectivity and voices of those studied.
- Comparison decontextualizes phenomena, plucking them out of their original context in order to set them alongside other things, and thus strips them of much of the meaning that they carried in their original setting.
- Comparisons do not allow us to make the sorts of generalizations that comparativists want to make because the patterns perceived are invented or imagined by the scholars.
- Comparisons between religions have been driven by a theological agenda that historically privileged some religions over others but today tends toward a generalized religious universalism inimical to the naturalistic or scientific study of religion.
- Comparisons between religions tend to be descriptive rather than explanatory and thus lack scientific value.

In the wake of these critiques, Paden and others have advocated a new, more chastened approach to comparison. Three key points have emerged:

- All comparisons are limited and aspectival. They can or should be structured around an "exact, stipulated point of analogy" (Paden 2005, 1880).
- The stipulated point of analogy structures the object of study and should be chosen in light of the question the scholar seeks to answer (Smith 1990).
- In stipulating a point of analogy between things compared, the scholar highlights an aspect of the phenomena. Comparisons thus connect two or more phenomena in order to explore how they *differ* in relation to the stipulated point of analogy. A stipulated point of analogy, in other words, provides a point of contact for exploring differences (Paden 2005, 1880).

In setting up comparisons, scholars of religion have typically rooted their comparisons in concepts—such as scripture, sacrifice, pilgrimage, or prophecy—that allowed them to compare one religious tradition to another and to avoid comparisons of things that some consider religious and others do not. Under the influence of the sui generis approach to religion, scholars assumed that the boundary between religious and non-religious things was fixed and stable, even if sometimes hard to discern. With a sui generis understanding of religion and religious experience there is no need to focus on this boundary to analyze how things become religious because religious things "just are." In light of these assumptions, scholars valued comparisons between religions and disparaged comparisons between religious and nonreligious things as reductionist. Where scholars of religion traditionally compared "religious things," an attributional approach takes as its object of study "*things* that may or may not be considered religious." Setting up an object of study thus requires us to specify the "thing" that we are going to study.

Although I have referred to "things" throughout the book, now, at the point of setting up comparisons, we have to be careful how we think about them. A few scholars of religion have proposed that we root our comparisons in "human universals" (Paden 2001, 276; 2005) or "a biological or cultural archive" (Martin 2004, 38–39). This approach would suggest that rather than comparing "religious things," we should take as the basis for comparison human universals (the "things" in this formulation) to which people sometimes ascribe religious characteristics. This approach, however, has certain limitations. First, it presupposes that we know what those universals are. In some cases (for example, birth, death, sleep) we do, in other cases (for example, paranormal experiences) it is not so clear. Second, it restricts our comparisons to things that are human universals. Third, as with the idea of "religious things," it perpetuates an overly reified sense of "things" (for example, universally human things).

We would be better off if we think of comparisons as premised on the identification of a *set of things* where the set in question is *temporarily constituted for the purposes of research by a stipulated point of analogy.* This has two advantages. First, it allows researchers maximum flexibility in terms of creating sets. To create a set, researchers simply have to identify a stipulated point of analogy that defines (and creates) a temporary set of things. We then can collect data about the things in the set (as discussed in chapter 2; see table 2.1). Since the accounts of subjects and observers are a crucial part of the data, the representation of the behavioral event on the ground, however contested, remains a defining feature of the "thing" even after it is incorporated into a set for the purposes of comparison. Second, this approach does not presuppose that we know where the boundary lies between the biological and the cultural or even that there is a clear boundary. Setting up comparisons in this way, however, does provide a way to determine how common certain types of experiences are across time and cultures and thus presumably the degree to which they are rooted in common biological processes or in processes that are more culturally specific.

Stipulating a point of analogy in order to constitute a temporary set of things is not as easy as it sounds, if we want to ensure that people's representations of things on the ground are not obscured in the creation of the set. Because things that share certain features in common may differ dramatically both in terms of their phenomenology and in the way that people represent them, we need to specify a point of analogy between things we want to study in terms that do not do violence to the disparate ways that people understand them. Although we might think that we could get around this problem by using academic categories, many such terms are loaded with disciplinary presuppositions that we also need to avoid. In other words, when comparing experiences considered religious by either subjects or scholars, the specification of a stipulated point of analogy can be complicated by disagreements at multiple levels over the nature of the experience in question. These disagreements are evident between subjects and observers in a particular cultural context, between people in different cultures and historical periods, and between researchers in different academic disciplines. At each level the assumptions embedded in concepts used to describe the experiences that interest us make it difficult to say what we are studying without taking sides in the disagreements, whether at the level of the primary data generated by subjects and observers or at the level of scholarship. We can consider both levels in turn.

Primary data.   When subjects and observers report on a subject's experience, the terms they use to describe the experience typically imply theories of causation that embed assessments of truth and authenticity (often involving strategies of discernment) that in turn have social and political implications. Thus, for example, a subject might describe her experience as inspiration, while critical observers might describe it as heresy or "enthusiasm" (an historical term for "false inspiration"). Another individual might claim to have seen an apparition of a holy person or a ghost; a critic might say the person imagined it. Although interactions between subjects and observers typically presuppose mutually intelligible cultural worldviews, assumptions can shift much more markedly if we seek to compare experiential claims made in one historical or cultural context with those made in another. Thus in one context it might be clear that our subjects are arguing over what constitutes authentic religion in relation to competing prophetic claims, while in another context those arguing about oracular pronouncements may have no easily discernible concept of either prophecy or religion. Nonetheless disputes over the authenticity of oracular pronouncements may strike the researcher as having certain features in common with disputes over prophetic claims, such that a point of analogy could be stipulated, despite that fact that one group refers to oracles and the other to prophets.

Disciplines.   Many concepts are claimed by disciplines and imbued with their disciplinary presuppositions. For example, experiences that scholars of religion might view as mysticism or ecstasy or prophecy might be viewed as shamanism or spirit possession by anthropologists and as symptoms of hysteria or dissociative disorders by psychiatrists. Many of our academic concepts are not only linked to disciplines but are laden with theoretical presuppositions peculiar to the discipline in question. Thus, for example, psychiatrists define hallucinations in terms of false perceptions of external data in light of their concern with voices and visions associated with psychological illnesses, and in so doing leave many to assume that all experiences that arise from internal perceptions are pathological. If we use such terms unreflectively, we may inadvertently reproduce long-standing historical controversies at an academic level, which in turn hinders analysis of the work that the experiences in question are doing (religiously, socially, and politically) in various cultural contexts.

Some of the difficulties surrounding academic or intellectual categories lie in the fact that they have emerged and been used within disciplines in which scholars are not in the habit of communicating with one another. As Cosmides, Tooby, and Barkow (1992) point out, there is little conceptual integration across disciplines outside the natural sciences.

> A compatibility principle is so taken for granted in the natural sciences that it is rarely articulated, although generally applied; the natural sciences are understood to be continuous. Such is not the case in the behavioral and social sciences. Evolutionary biology, psychology, psychiatry, anthropology, sociology, history, and economics [not to mention religion] largely live in inglorious isolation from one another: Unlike the natural sciences, training in one of these fields does not regularly entail a shared understanding of the fundamentals of the others (Barkow, Cosmides, and Tooby 1992, 4).[1]

Disciplinary isolation and the resultant lack of conceptual integration or even testing of concepts across disciplinary boundaries has seriously hampered the construction of comparative objects of study that make sense across disciplinary lines.

Despite these difficulties, there is no escaping the use of categories in setting up comparisons. The crucial question is not *whether* we are going to utilize categories in order to compare but *what* categories we are going to

---

[1] Barkow, Cosmides, and Tooby (1992) define "conceptual integration—also known as vertical integration— . . . [as] the principle that the various disciplines within the behavioral and social sciences should make themselves mutually consistent, and consistent with what is known in the natural sciences as well. The natural sciences are already mutually consistent; the laws of chemistry are compatible with the laws of physics, even though they are not reducible to them. . . . A conceptually integrated theory is one framed so that it is compatible with data and theory from other relevant fields" (4).

use. While the phenomenological description of the subject's experience should reflect the terms used by the subject and indigenous observers, the way that researchers compare these experiences, and thus the categories that are utilized in the act of comparing them, are the responsibility of the researcher (Satlow 2005, 287–98).

SPECIFYING A POINT OF COMPARISON

The primary challenge that scholars face in setting up comparisons of experiences is that of specifying a point of analogy between experiences that subjects view as very disparate in ways that do not unduly violate subjects' sensibilities. In other words, our aim should be to specify a point of comparison that allows us to engage experiential phenomena across traditions of interpretation without unduly violating the lived experience of those within them. Insofar as possible, we should avoid what Proudfoot (1985, 180–81) has termed "descriptive reduction" by specifying the experiences in question "under a description that can plausibly be ascribed to the person to whom we attribute the experience." Doing so usually involves moving to a higher level of abstraction or generality. Oftentimes there is no ready-made concept that works for all parties we want to compare. In such cases we can use extended descriptive statements instead.

Ethnographers and experimentalists can elicit reactions to concepts or descriptive statements directly to see if they have arrived at a sufficiently generic formulation. Historians must be familiar enough with the writings of their historical subjects to test their formulations against those of their subjects. For example, in *Fits, Trances, and Visions* (1999), I was interested in subjects whose usual sense of themselves was altered or discontinuous, including experiences that involved the loss of motor control, altered sensory experiences, and discontinuities of consciousness, memory, or identity. I tested this formulation against the writings of Jonathan Edwards, one of the most intellectually rigorous figures discussed in the book. Because Edwards understood authentic religious experience as involving a "new supernatural sense" entirely distinct from the natural senses, I changed *altered* (and thus presumably still natural) sensory perceptions to *unusual* sensory experiences (that presumably could be understood as supernatural). This minor change accommodated his perspective without undercutting my goals in setting up the comparison.

Researchers can also test out concepts across disciplines, redefining terms or developing descriptive statements in cases in which disciplinary terminology is incompatible or too theoretically laden. In writing *Fits, Trances, and Visions*, it took a while to arrive at a relatively neutral way to describe the set of experiences variously designated as "dissociation" in psychiatry; "trance," "spirit possession," and "altered states of consciousness" in anthropology; and "visions," "inspiration," "mysticism," and "ecstasy" in religious studies. Adopting any one of these terms risked positioning the work in relation to disciplinary subject matters (for example, religion, culture, or mental illness) and explanatory commitments (religious, cultural, or pathological) to which I was not committed.

In the end I turned to the sections on conversion and dissociative disorders in the American Psychiatric Association's *Diagnostic and Statistical Manual* (DSM-IV) for inspiration. While this might seem an unlikely source under the circumstances, the DSM seeks to formulate its descriptive terminology as neutrally as possible in order to avoid invoking any particular psychiatric explanation for a diagnosis. Its care in this regard made it a helpful starting point for constructing a generic description of the experiences that interested me. Because, as the DSM takes care to acknowledge, the psychiatric labels apply only in situations of mental distress, I did not use them in the book. In this case, extended descriptive statements thus provided a way to negotiate contestations between disciplines as well as between subjects. They can also be used to set up comparisons in which the academic categories are part of the subject matter being studied.

Before moving on to distinguish between different types of comparisons that we can make, let us sum up this section by considering the relationship between the set of things constituted by a stipulated point of analogy and the various kinds of special things depicted in figure 1.1, p. 45, again using *Fits, Trances, and Visions* as an example. In stipulating alterations in people's usual sense of themselves as a point of analogy between experiences, I created a set of things, in this case experiences, that were by definition fairly unusual and thus likely to stand out for people as special in many cultural contexts. As such, they would most likely fall within the limits of the box in figure 1.1 and more specifically within the circles that include anomalous experiences.

The stipulated point of analogy that defined the set of things in this particular case just happened to fall largely within the set of things that people consider special. Dreams provide a more complicated example. Depending on how researchers want to define "usual," dreams can be defined in or out of the data set defined by alterations in people's usual sense of themselves. If they associate "usual" with waking consciousness, dreams would be in the data set. If the data set is defined in terms of *unusual alterations* in people's sense of themselves, dreams, understood as usual alterations, would be excluded. Whether or not dreams are defined as part of the data set, most dreams would not fall into the set of special things, since most people consider most of their dreams pretty ordinary. People do consider some dreams as special, however, and people in cultures that

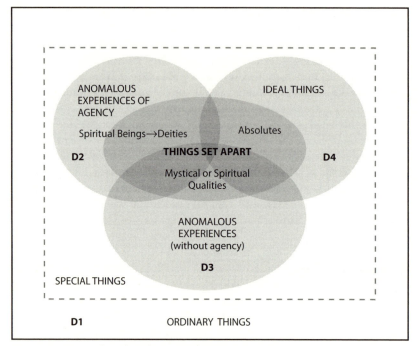

Figure 4.1. Examples of Ordinary and Special Dreams

cultivate dreams and look to them for insight might have more dreams they consider special than people in cultures that do not. If dreams are considered special, they can be special in any of the ways specified in figure 1.1. We can illustrate this by locating four different types of dreams (an ordinary dream [D1] and special dreams involving an ancestor [D2], an out-of-body experience [D3], and a feeling of complete happiness [D4]) in a modified version of figure 1.1 (figure 4.1).

To summarize, researchers constitute sets of things for research purposes by stipulating a point of analogy between them. These sets of things may or may not include special things. If researchers want to investigate the conditions under which people remember very ordinary dreams, they would include D1 in their data set but not dreams D2–D4. If they were interested in intense feelings of happiness, they would include D4 along with intense feelings of happiness that occur while people are awake. The key point to remember is that *researchers* constitute sets of things for research purposes, while *subjects* decide what things are special through context-specific assessments of value. In this chapter we will examine various ways that researchers can constitute sets of things in order to examine how subjects assess value under different sorts of conditions.

TABLE 4.1
Types of Comparisons

Type of Comparison	Intracontextual	Intercontextual
Simple and Simple	Conflicts over interpretation. Is it really as you say it is?	Identification of cross-culturally stable forms of experience
Composite and Composite	(1) Conflict over efficacy of practices relative to goals. Does it really do it? (2) Conflict over goals. Is it the right goal?	(1) Identification of processes by means of which practices effect goals (2) Identification of ends/ goals of paths.
Simple (SF) and Composite (CF)	Conflicts over legitimacy of SF relative to CF and legitimacy of CF in light of SF. Is it authentic?	Analysis of processes of making and unmaking religions

COMPARING SIMPLE AND COMPOSITE FORMATIONS

In setting up comparisons we can distinguish between those that compare things in one context or connected contexts and those that compare things in contexts that are disconnected in space and/or time. In the former, which we can call "intracontextual" comparisons, subjects as well as researchers may recognize points of analogy between the things compared and have investments of their own in the comparison; in the latter, which we can call "intercontextual" comparisons, the subjects in one context are unaware of subjects in the other and the point of analogy between the two contexts is noticed only by the scholar. Both ways of setting up comparisons allow scholars to get at interesting questions and pose different challenges with respect to stipulating a point of comparison. We can use this distinction to explain the kinds of questions that we can explore when we set up comparisons between two or more simple ascriptions, two or more composites ascriptions, and simple and composite ascriptions (table 4.1).

Simple and simple. In comparisons of this sort, researchers compare simple ascriptive formations, which they constitute as a set for research purposes on the basis of a common feature or set of features that they identify. Those who have the experiences and those who observe them may consider some of the experiences in the set as ordinary and others as special to varying degrees and will not necessarily agree with one another.

Intracontextual comparisons of this sort allow us to focus on conflicts over the interpretation of the experiences in a particular historical or cultural context. The primary questions that experiencing subjects and those around them ask are: How do I/we explain this experience? Is it really what you say it is?

Intercontextual comparisons allow us to assess the cross-cultural stability of various forms of experience. Identifying cross-culturally stable forms of experience is sometimes fairly straightforward, as, for example, in the case of dreams, auditory or visual hallucinations, or out-of-body experiences. In other cases, such as sleep paralysis or parapsychological phenomena, identifying an underlying cross-culturally stable form of experience is a major preliminary step in its own right. At this level, researchers can ask: Are there unusual experiences that are nonetheless cross-culturally stable and grounded in potentially identifiable, albeit unusual, biological processes? Do people tend to consider these experiences special? If so, what features cause them to stand out?

Composite and composite. In comparisons of this sort, we can construct an object of study in which specified ascriptions, constitutive of the composite, form the stipulated point of analogy between composites. Using the special-paths composite, we can compare paths that share a common goal (for example, participating in the real presence of Christ) but differ with respect to the practices deemed efficacious for attaining the goal. Or we can compare paths that utilize common practices (for example, prayer practices) to achieve very different goals. When the goal is held constant, an intracontextual comparison allows us to focus on debates within a tradition or family of traditions over the efficacy of practices relative to the common goal. If the goal is allowed to vary, intracontextual comparisons allow us to focus on debates between traditions over the value of the goal itself. The primary questions subjects ask in this type of comparison are: Do the practices really effect the goal? Is the goal the right one? If we hold the type of practice constant in an intercontextual comparison, we can compare how the practice can be directed to different ends by different traditions and thus how the practice works to effect various goals. At this level, the primary question is: Can we (as scholars) identify cross-culturally stable processes by means of which practices effect goals? If practices rely on common underlying processes, does this lead to corresponding similarities in the way the goals are understood?

Simple and composite. In comparisons of this sort, we can construct an object of study in which a feature common to both the simple and the composite formation forms the stipulated point of analogy. Intracontextualized comparisons of this sort allow us to focus on conflicts over the legitimacy of simple formations (what individuals claim they experience) relative to composite formations (whether existing or emergent) and the challenges that simple formations pose to existing composite formations. The primary question for subjects is: Is this experience authentic? Intercontextual comparisons of this sort allow us to compare the processes by which, on the one hand, individual experience is modified and controlled in light of the composite and, on the other, new religions emerge and existing ones are reformed. The primary question for researchers at this level is: what is the role of individual experience in the making and unmaking of religions?

Simple and Simple

Intracontextual comparisons of simple ascriptions—experiences that are sometimes considered religious and sometimes not—require that we stipulate the specific points of analogy between the experiences we want to compare. This ordinarily means specifying the common features of the experiences in question in terms that do not override subjects' interpretations of the experience. In the example given above, the point of analogy— experiences in which subjects' usual sense of themselves was altered or discontinuous—was formulated for an intracontextual comparison. This specification allowed me to track conflicting interpretations of experiences that some Anglo-Americans characterized as religious (most typically in that context in terms of the power or presence or indwelling of God, Christ, the Holy Spirit, or spirits) and by others as nonreligious (in terms of the imagination, animal magnetism, mesmerism, hysteria, hypnosis, subconscious automatisms, and suggestion) over a broad swath of historical time (Taves 1999, 3).

In contrast, intercontextual comparisons of simple ascriptions can be used to investigate the cross-cultural stability of various forms of experience. The research on sleep paralysis, which has been recognized as a cross-culturally stable experience relatively recently, provides a model for investigating other potentially stable complex experiences. For our purposes, we can characterize this research as unfolding in three phases. In the first phase, researchers identified a cross-culturally stable form of experience—now designated by the second-order term "sleep paralysis"— embedded in first-order narratives from various cultures and time periods. In the second phase, which is still in progress, they are working to characterize and explain the various sorts of experiential phenomena (for example, felt presence, pressure on the chest, and out-of-body experiences) associated with sleep paralysis. In a potential third phase, we could set up comparisons that would allow us investigate the conditions under which experiences that manifest the core features of sleep paralysis are considered religious or not.

Phase 1.   Much of the credit for identifying sleep paralysis as a cross-culturally stable form of experience goes to folklorist David Hufford. Hufford's (1982) research began with the "Old Hag tradition" in Newfoundland, which he discovered while teaching and researching folklore at the Memorial University of Newfoundland in the early 1970s. The tradition recounted incidents in which an "old hag" attacked people at night. The victims, who claimed to be fully awake, were unable to move to defend themselves against the attacks, which they found terrifying.

Hufford began his research on the tradition by distributing a questionnaire to students asking for more information on the "nightmare/hag/old hag." He used all three terms because he understood them to be related, though he was not clear about the nature of the relationship at the outset (Hufford 1982, 3). Students were asked to write down detailed descriptions of their own experiences or those of others, especially older persons. Hufford supplied a list of questions about the experience, indicating that, though he used the word "hag" throughout, students were to "use the word which is known in [their] community" (Hufford 1982, 257–58). Based on these initial accounts, Hufford characterized the experience as one in which a person either during or immediately preceding sleep heard or saw something come into the room that pressed them on the chest or strangled them. The person was unable to move or cry out during the attack.

In order to determine whether this experience was local and culturally specific or cross-culturally stable, Hufford identified the features that he thought might comprise an underlying cross-culturally stable experience and developed a questionnaire to test his hypothesis. The questionnaire asked respondents if they had "ever been awakened during the night to find [themselves] paralyzed; that is, unable to move or cry out?" Only after a series of questions requesting more information about the experience did he ask if they had a name or names for it (17–18). Although, due to various constraints, he formally administered the questionnaire only to students in Newfoundland, he carried out more informal surveys in mainland Canada and the United States (50). Based on both in-depth interviews and cross-cultural survey data, Hufford concluded that what Newfoundlanders called the "Old Hag" was an experience with a distinct, complex, yet stable pattern that could be found in other cultural settings. Based on a review of the medical literature, Hufford concluded that the old hag experience fit under the general heading of sleep paralysis with hypnagogic and hypnopompic hallucinations—that is, a brief experience of involuntary immobility immediately prior to falling asleep (hypnagogic) or upon waking (hypnopompic), often associated with vivid hallucinations of a sensed presence.

Using historical and ethnographic methods, researchers can identify complex experiences that are cross-culturally stable. They can also identify questions that experimental researchers need to address. Although there was a growing body of research on sleep paralysis, Hufford pointed out that researchers had no explanation for the remarkable stability of the content—that is, the frequency with which persons reported feeling assaulted by a felt presence (1982, 149–70, 245–46). Experimentalists subsequently took up this question.

Phase 2. Sleep researchers now know considerably more about sleep paralysis than they did when Hufford wrote his book. Neurological data suggests that sleep paralysis is the result of the intrusion of the REM (rapid eye movement) state into consciousness during transitions between sleeping and waking (Hishikawa and Shimizu 1995). Sleep paralysis episodes combine dream images and feelings with waking consciousness to create a waking dream state. Large-scale questionnaires have further refined the types of hallucinations that occur in conjunction with sleep paralysis. Based on data derived from a large-scale retrospective questionnaire, Cheyne et al. (Cheyne, Rueffer, and Newby-Clark 1999; Cheyne 2003) proposed a three-factor structure for categorizing the hallucinations associated with sleep paralysis, differentiating between experiences consistent with threatening intruders (Intruder), physical assaults (Incubus), and vestibular-motor sensations (V-M). The first category includes experiences of a sensed presence, fear, and auditory and visual hallucinations. The second includes sensations of pressure on the chest, difficulty breathing, and pain. The last consists of floating or flying sensations, out-of-body-experiences, and feelings of bliss.

Cheyne and Girard (2007a) were able to test and refine their categories using more immediate post hoc data generated by means of a Web-based questionnaire. In a recent study, which included 383 subjects, visual or auditory hallucinations of a presence were the most common (58 percent). Tactile hallucinations (linked to a sensation of pressure [53 percent] or breathing difficulties [47 percent]) were slightly less common, while vestibular-motor hallucinations were the least common (31 percent). In all categories, the experiences were most often perceived as fearful. In contrast to earlier data, this extended to the V-M experiences as well, with 85 percent of V-M experiences associated with fear and only 12 percent associated with bliss.

The use of Web-based questionnaires provides a promising vehicle for the collection of immediate post hoc data on unusual experiences, such as sleep paralysis, from a diverse subject population. Sleep paralysis data, like dream data, are all post hoc. Immediate post hoc sleep paralysis data are more difficult to collect, however, because sleep paralysis episodes occur unpredictably, even though they are not uncommon in the general population (estimates range from 6–40 percent). In traditional reporting procedures, subjects typically were asked to report on all past occurrences. This meant that the time elapsed between the experience and the report varied widely and that different occurrences were easily conflated. Using a Web-based questionnaire, however, Cheyne and Girard (2007a) gathered reports from subjects immediately after the experience, and in come cases they were able to accumulate multiple immediate post hoc accounts from the same subject. The use of the Web also allowed them to expand their subject pool beyond the usual undergraduate population and included subjects from around the world, though the United States (65 percent), Great Britain (11 percent), and Canada (8 percent) were significantly over-represented.

Cheyne and Girard (2007a) hypothesized that Intruder and Incubus experiences, both of which involve a felt presence, constitute a distinct cluster of experiences that could be empirically distinguished from V-M experiences that involve alterations in the body-self neuromatrix. As anticipated, they found that immediate post hoc reports, which avoid the conflation of multiple experiences, showed greater independence between accounts of Intruder and Incubus experiences, on the one hand, and V-M experiences, on the other. Their data showed a clear distinction between sleep paralysis experiences that involved a felt presence and experiences that involved altered relationships between self and body (2007b, 4).

Researchers are now debating two competing explanations of the felt-presence experiences associated with sleep paralysis. Cheyne (2001; Cheyne and Girard 2007a) theorizes that they result from pre-hallucinatory activation of the threat-vigilance system. Alternatively, Nielsen (Nielsen 2007; Solmonova et al. 2007) theorizes that a paralysis attack activates hallucinatory social imagery primed by REM sleep processes resulting in a sense of felt presence, especially in persons prone to anxiety disorders or affect distress. In their response to Nielson, Cheyne and Girard (2007b) agree that sleep paralysis experiences most likely do utilize social imagery, but they argue that this does not explain the biases in the imagery generated or the evidence that the felt presence arises prior to sensory experiences. According to Cheyne and Girard (2007b, 2),

> The essence of FP [felt presence] is that it is the experience . . . of a *feeling* of the presence of a *Being*, and not as a mere existent thing, but an *intentional* (Dennett 1987) Being with a mind or soul. . . . It is clear from the very explicit description of FP experients (Cheyne 2001) that the FP as typically experienced during SP [sleep paralysis] is best described as an intuitive "gut feeling" (Damasio 1994; 1999) or what Woody and Szechtman (2000) call "yadasentience."

They theorize that the experience of feeling awake, typically while alone and in the dark, and yet unable to move triggers the threat-activation

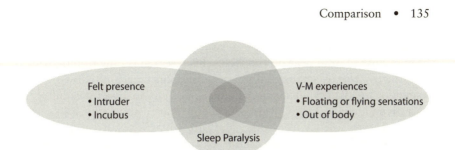

Figure 4.2. Two Clusters of Experiences Associated with Sleep Paralysis

system (TAVS). This system, which is similar to what some cognitive scientists of religion (for example, Barrett 2004) refer to as the hyperactive agency detection device (HADD), has an evolved bias toward the detection of agents in ambiguous situations (Cheyne and Girard 2007b, 6). The experience of felt presence in sleep paralysis thus provides a context in which it may be possible to test theories that link agency detection with the construction of unseen agents.

Although sleep paralysis is relatively well defined, the experiences associated with it fall into two broad clusters—felt presence and vestibular-motor—that are less well defined and also occur outside the context of sleep paralysis (figure 4.2). Based on the current state of research we cannot assume that experiences that fall within each of these clusters have a common neurological basis. Thus, although Nielson (2007) provides a long list of contexts in which researchers have reported a sense of felt presence, including epilepsy, brain damage, sensory deprivation, extreme environmental conditions, and bereavement, Cheyne and Girard (2007b) caution against conflating these felt-presence experiences and assuming that common cognitive processes mediate them all. The situation is similar with V-M experiences—that is, floating or flying sensations and out-of-body-experiences—which also can occur outside the context of sleep paralysis. They, too, are the subject of intensive research, and here, too, researchers are just beginning to sort out the different kinds of anomalous experiences of the self based on phenomenal and neuroscientific data (see, for example, Blanke and Mohr 2005 and Arzy et al. 2006). Research on the two clusters of experiences promises to provide new insight into the neurological basis of anomalous feelings of presence that may be attributed to the presence of invisible agents and anomalous experiences of the self that are often characterized as "mystical" or "spiritual."

Phase 3. Psychological and neuroscientific research on sleep paralysis and associated experiences is refining our understanding of the cross-culturally stable features of these experiences and generating thousands of narrative accounts of sleep paralysis experiences, some of which have been published or appear on the Web. This research is now feeding back into the analysis of ethnographic and historical data and allowing researchers to identify incidents of sleep paralysis in disparate cultural narratives, including accounts of sexual abuse, alien abduction, and near-death experiences (Clancy 2005; Hufford 2005; McNally and Clancy 2005a, b; Nelson et al. 2006). Hufford noted the particular ease with which individual Mormons were able to interpret sleep paralysis experiences religiously due to their tradition's openness to the supernatural, and suggests that a similar openness may exist among charismatic and Pentecostal groups (1982, 222–25).

Greater awareness of sleep paralysis research has not resulted in the straightforward secularization of the experience among those affected. In keeping with the general tendency of esoteric or "new age" movements to appropriate psychological research (Hanegraaff 1998, 224–28, 482–513), some people are using sleep paralysis experiences as a gateway to esoteric realms. The website of the Trionic Research Institute provides an explicit bridge between the scientific and the esoteric realms (Hufford 2005, 30–32). Acknowledging that sleep paralysis experiences can be terrifying, the website nonetheless proclaims, "beyond the fear there is a gate." The website explains that experiencers can either "shut the gate and walk away" or "walk through the gate" in order to transform the sleep paralysis experience into "a pleasant lucid dream or Out-of-Body Experience." The site provides links to experiencer sites, research sites, accounts of sleep paralysis in literature, and discussion of sleep paralysis in relation to out-of-body and alien-abduction experiences. The "Explorations in Consciousness" website, which is devoted to out-of-body experiences, lucid dreams, and altered states of consciousness, makes no specific mention of sleep paralysis, but provides information on out-of-body experiences and lucid dreaming from scientific, occult, and "contemporary" perspectives. The contemporary perspectives include an extensive list of works by nonscientific explorers of out-of-body experiences, past-life regression, and near-death experiences.[2]

Research on sleep paralysis has important implications for the study of religion. The chief feature of sleep paralysis experiences, especially those with accompanying felt presence or out-of-body experiences, is how *real* the hallucinations of felt presences or out-of-body experiences feel to those who are paralyzed. Like Barnard's V-M experience, these hallucinations occur spontaneously and usually do not reflect the preconceived ideas of the subject in any strong sense. Hufford (1982, 2005) in particular stresses

---

[2] Trionica Research Institute, available online at http://www.trionica.com/; Explorations in Consciousness, available online at http://www.explorations-in-consciousness.com/index.htm.

this point. Both Cheyne and Hufford note subjects' tendency to use religious language to describe the sense of felt presence even if they do not think of themselves as religious. Thus many make spontaneous references to anomalous agents (for example, devils, ghosts, aliens, gods) even if they do not believe such agents exist. Cheyne also stresses that whether or not subjects actually believe they left their bodies, they still feel as if they did (Cheyne 2002). The experiences associated with sleep paralysis thus provide a quintessential example of experiences that *seem* sui generis.

As a result of this feeling of realness and the growing awareness of scientific research on sleep paralysis, Cheyne finds many in his recent samples who operate on two levels of interpretation with respect to the experiences that accompany sleep paralysis.

> The first-level interpretations are on-line, situationally embedded, automatic, intuitive and obligate [that is, first-order responses]. One might describe this level of consciousness as "the way things seem." There is, however, a second level of interpretation that is more explicit, critical and verbal [that is, higher-order]. This level of interpretation is an off-line or "action-neutral" . . . decision about the first-level experiences and may be described in everyday terms as "the way things 'really' (or probably) are." As the parenthetical "probably" implies these higher-order interpretations are often circumspect and tentative (Cheyne 2001, 147).

This is an important distinction that approximates the distinction that cognitive scientists of religion have made between "cognitively optimal" and more reflective (and thus more "cognitively costly") beliefs (Barrett 2004, 31–43, 107–18; Whitehouse 2004, 29–33). Sleep paralysis experiences, in other words, provide an excellent example of experiences that people "naturally" interpret in a certain way, even if upon reflection they believe that their spontaneous interpretations were inaccurate.

A considerable body of experimental research indicates that people readily ascribe agency to perceptions based on very sketchy information and then make inferences about the "agents'" behavior based on assumptions (theory of mind) that we apply to agents more generally. This happens particularly in situations of ambiguity or threat (see citations in Barrett 2004, 32–34). Barrett and Keil (1996) found that subjects possessed and used at least two different parallel god concepts, an anthropomorphic concept that they drew upon unreflectively in contexts requiring relatively automatic or on-line processing (in this case interpreting stories) and a more theologically complex concept that they drew upon when asked to reflect. Combined with the human tendency to over-attribute agency in situations of ambiguity or threat, this distinction between relatively automatic and reflective god concepts helps to explain why people may spontaneously refer to agents that under other circumstances they would say did not exist. Cheyne and Girard (2007b) explicitly connect their theory that felt-presence experiences are threat-activated with work on hyperactive agency detection. This suggests that sleep paralysis specifies both a situation in which the seeming reality of anomalous agents is heightened and a context that may be responsible for generating a disproportionate share of the most vivid reports of encounters with such agents.

Discussions of cognitively optimal religion have been premised entirely on religion defined in terms of counterintuitive agents. We need further research to see if there is a natural tendency to ascribe significance to unusual vestibular-motor experiences. When processing them on-line, do people draw on other aspects of folk psychology—say, an innate tendency toward body-self dualism—to set such experiences apart as special? Although the widespread significance given to out-of-body and other vestibular-motor experiences, such as Barnard's, makes this seem plausible, this idea needs to be tested.

Both Hufford and Cheyne believe that research on sleep paralysis has important implications for understanding religious or spiritual beliefs and draw on various classical theorists to make their case. Cheyne offers a naturalistic interpretation of Rudolf Otto's idea of the "ominous numinous" in light of sleep paralysis experiences, while Hufford offers a neo-Tylorian view of the origins of religion grounded in the belief in spirits in light of the same evidence (2005, 14). The text most often cited, however, is the passage in James's *Varieties of Religious Experience* in which James states: "It is as if there were in the human consciousness *a sense of reality, a feeling of objective presence, a perception* of what we may call '*something there*,' more deep and more general than any of the special and particular 'senses'" (quoted in Cheyne 2001, 133; Hufford 2005, 15 [emphasis in original]).

This "sense of reality" that is deeper and more general than "the special and particular 'senses'" is central to James's overall argument; it is introduced in the passage quoted above and elaborated in James's discussion of the sense or consciousness of the "More." The evidence James offers for this sense or feeling sounds remarkably like a case of sleep paralysis. He indicates that "curious proofs of the existence of such an undifferentiated sense of reality . . . are found in experiences of hallucination" in which "the person affected will feel a 'presence' in the room, definitely localized, facing in one particular way, real in the most emphatic sense of the word, often coming suddenly, and as suddenly gone; and yet neither seen, heard, touched, nor cognized in any of the usual 'sensible' ways" (James 1902/1985, 55). He cites the example of a friend—the psychical researcher Richard Hodgson—who had several experiences of this sort.

Hodgson described three of these experiences in some detail in letters to James written at James's request some years after the experiences occurred. All the experiences occurred soon after Hodgson went to bed. The first involved a "vivid tactile hallucination of being grasped by the arm," the second and third involved the feeling of "something com[ing] into the room and stay[ing] close to [his] bed" along with "a very large tearing vital pain spreading chiefly over the chest," accompanied by a feeling of "abhorrence" (James 1902/1985).[3] Although Hodgson evidently did not experience paralysis, his account otherwise conforms to that of sleep paralysis of the Intruder-Incubus type.

Lacking a naturalistic explanation for such seemingly real experiences of a felt presence, scholars such as Proudfoot and Barnard alternatively over- and underinterpreted these experiences. Proudfoot minimized the experiences, arguing that they have "the feel of direct sensations, but . . . the epistemic status of hypotheses"—that is, hunches or guesses (Proudfoot 1985, 161–64). Barnard rightly claimed that Proudfoot did not take Hodgson's description seriously enough. Barnard, however, overinterpreted Hodgson's sense of the reality of his hallucinations, using it as evidence for his (Barnard's) inferential claims about the reality of mystical experience (Barnard 1997, 107–10). Hodgson himself interpreted his experiences more cautiously, emphasizing the intensity and certainty of his *feeling* that there was an actual presence in the room with him.

The idea of a felt presence also appears in Rudolf Otto's *Idea of the Holy*, another classical theory of religion. Otto quotes James's "feeling of objective presence" approvingly and recasts it as his signature "feeling of a 'numinous' object" (Otto 1950, 11), one of the two types of experiences most commonly viewed as inherently religious. While Cheyne's naturalistic interpretation of Otto's "ominous numinous" in light of sleep paralysis experiences is interesting, all it really tells us is that some who experience a felt presence in the context of sleep paralysis find language such as Otto's helpful in describing their experiences.

Theorizing about experiences people consider religious along the lines suggested by James, Barnard, Hufford, and Cheyne strikes me as premature. In time we may be able to identify a variety of *types* of experiences that—empirically speaking—people are more likely to consider special and to which, as a result, they are more likely to attach religious ascriptions across cultures than would be the case with other sorts of experience. At this point, we need less grand theory building and more careful

[3] For the correspondence about these experiences between James and Hodgson, see James 1992–2004, 9:291–92, 609. When James asked about Hodgson's reference to "a Mighty Being," Hodgson indicated that he experienced a "bigger personality" than his own, not God, "though in the way to God" (609).

empirical studies that will allow us to identify unusual experiences that are relatively stable across cultures. Hufford's method, as he stressed, has potential in this regard in so far as it allows us to assess the extent to which any "apparently fantastic beliefs are in fact empirically grounded" in cross-culturally stable, complex experiences. He suggested in the early 1980s that near-death experiences, mystical experiences, and various sorts of "paranormal" phenomena were ripe for such reexamination (Hufford 1982, 245–56; Foster 1985, 168–74) and much of this work still remains to be done (see in this regard McClenon 1994; Sherwood 2002).

Composite and Composite

Experience plays a different role in composite formations than it does in simple ones. Special paths, defined in terms of practices that followers of the path believe are efficacious for achieving the goals specified by the path, do not necessarily even mention experience. In so far as the goal of the path is to re-present an original or prototypical event to the followers of the path, the goal of the path is likely to be formulated in experiential terms (for example, experiencing the presence of Christ or enlightenment). Discussions of such goals typically center on normative expectations related directly or indirectly to the re-creation of the originary event—that is, an event deemed religious (Jesus deemed the Messiah [the Christ] or Gautama deemed the Enlightened One [the Buddha])—rather than on first-person accounts of immediate sensory or perceptual experiences. If descriptions of experience that seem to be perceptually based are included in such discussions, their inclusion typically reflects normative concerns and illustrates prescriptive expectations. In short, discussions *about* the experiential goal of a path are not the same as experiences *of* the goal of a path. The former are specified formally—that is, in an official, prescriptive sense that is often realized ritually. The latter vary widely. In this section, I will consider comparisons between paths conceived in this formal sense and defer discussion of the interplay between the normative expectations of the path and the actual experience of individuals or groups.

Comparisons of paths deemed religious can be set up in one of two ways: either by stipulating a point of analogy in terms of the practices deemed efficacious relative to the goal, or in terms of the goal itself. Intercontextual comparisons in which the practice forms the point of analogy allow scholars to compare how similar practices can be directed toward different ends in different traditions. Thus, for example, in *Veda and Torah*, Holdrege (1995) compares practices of transmission, interpretation, and appropriation of Vedic and Jewish texts that embody distinctive mythic understandings of the Word and are deemed efficacious relative to differently conceived religious goals. Comparisons of this sort can also

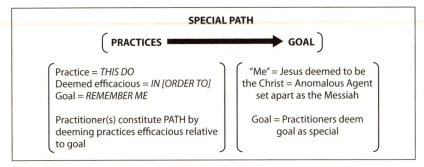

SPECIAL PATH

(PRACTICES ━━━━━━➤ GOAL)

Practice = *THIS DO*
Deemed efficacious = *IN [ORDER TO]*
Goal = *REMEMBER ME*

Practitioner(s) constitute PATH by
deeming practices efficacious relative
to goal

"Me" = Jesus deemed to be
the Christ = Anomalous Agent
set apart as the Messiah

Goal = Practitioners deem
goal as special

Figure 4.3. "This Do in Remembrance of Me" as Composite Ascription

be used to refine second-order concepts such as ritual. Thus Sharf (2005, 246–49) compares the Catholic mass and Buddhist ancestral offerings, using the comparison to identify common features of ritual despite the very different goals toward which the rituals are directed. Intercontextual comparisons of composites allow scholars to refine second-order concepts and to identify cross-culturally stable processes by means of which practices effect goals deemed religious.

Intracontextual comparisons of paths deemed religious, in which the goal forms the point of analogy, allow us to focus on debates over the efficacy of practices for arriving at the common goal. Wandel's (2006) study of the conflicts over the real presence of Christ in the Eucharist during the Reformation era provides an excellent illustration of the way that comparisons of this sort can be set up. Her book examines the sixteenth-century controversies over biblical passages in which Jesus refers to bread and wine as his body and blood. These passages, which are recited in virtually all Christian eucharistic rituals, form the link between the disciples' originary experience of Jesus and the re-presentation of the originary experience to followers in other times and places. Sixteenth-century controversies over the meaning of these words led to the formation of three distinct churches with liturgies that united followers (designated as Lutheran, Reformed, and Catholic) across linguistic and political boundaries.

Two biblical passages (Matthew 26:26–28 and Luke 22:19–20) constituted Wandel's point of comparison between contending Reformation-era figures and emergent groups. The two most contentious phrases were "this is my body" and "this do in remembrance of me." The second of the two phrases, "this do in remembrance of me," defines a path deemed religious, arguably the central formulation of the path for most Christian groups. Thus if we recast "in remembrance of me" as "in order to remember me," then "this do" is the practice, "remembering me" is the goal, and "in order to" links the practice with the goal, deeming it efficacious (figure 4.3).

Each word invoked many contentious issues in the context of Reformation-era debates:

- *"This"* was understood as having to do with the "taking and eating," which in turn was linked to what was eaten (the bread), which in turn was linked to what Jesus said about the bread—that is, "this is my body." (The drinking of the wine and its relation to "this is my blood" was also part of "this" but played a subsidiary role in the debates over what it meant to "do this.")
- *"Do"* was understood as a reference to the specific words and actions of the celebrant relative to the bread and wine, and by extension the ceremony as a whole, and the way the relationship between the celebrant and the congregation is constituted by the ceremony. (The specific words spoken by the celebrant were, of course, the scriptural words whose meaning the factions were debating.) Disputed issues included the nature of the celebrant (whether priest or minister), the status of the ritual that constituted and authorized the celebrant to "do this" (that is, whether ordination was a sacrament or not), and what exactly the celebrant *did* (effected) in the "doing" of "this."
- *"In [order to]"* linked the "doing of this" with the goal of "remembering me." In linking the two, it signaled that "doing this" would effectively accomplish the goal of "remembering me." This structural link, as such, was not disputed.
- *"Remember me"* (literally *commemorationem* [Vulgate], remembrance [King James]) was understood as a reference to the goal or purpose of "doing this." "Me" was understood to refer to Jesus, whom all parties to the debates understood as the Christ (that is, the Messiah). All parties to the disputes in question thus agreed that it was Christ who was "remembered" by "doing this." They all even agreed that remembering made Christ present in some sense; it was the sense in which *doing this made Christ present*—whether "really" or simply in memory, whether bodily or spiritually—that was at the center of the dispute.

The sixteenth-century controversies that led to the emergence of the Lutheran, Reformed, and Catholic understandings of the Eucharist can be formulated in terms of disagreements over the efficacy of practices relative to the goal of re-presenting Christ. Relative to the two basic ascriptions that make up the composite, disputants agreed that Jesus the human being should be deemed the Messiah/Christ, which meant, in light of the early creeds, that he was God (thus, constituting him as a counterintuitive agent), while disagreeing over which practices should be deemed efficacious for remembering or representing him.

Comparisons set up in this way allow us to investigate a number of important issues, including: (1) who does the deeming and who has the power to do so authoritatively; (2) the processes whereby some practices come to define translocal traditions, while others remain local or simply fade into obscurity; and (3) how different understandings of efficacy result in different normative expectations regarding experience. The following excursus uses the eucharistic controversies to illustrate these issues in more detail for those who are interested. For the purposes of the argument, it is sufficient to note that different understandings of ritual efficacy combined with different assumptions regarding how Christ could be present to the believer to create very different normative expectations regarding experience within the three traditions. As a result of these debates over efficacy, Protestants formulated new distinctions between magic and religion and ritual and experience that reflected their normative presuppositions. These distinctions, which continue to inform the academic study of religion, will be taken up following the excursus.

Excursus on Deeming Practices Efficacious: Reformation-era Controversies

On deeming, authoritative and otherwise. In a chapter on debates over the Eucharist in the free imperial city of Augsburg, Wandel (2006) demonstrates not only the diversity of views that emerged regarding efficacious practice but also the wide range of people who advanced them. Between 1518, when the first evangelical preachers arrived, and 1537, when the city council authorized a common eucharistic liturgy, the issue was debated throughout the town in a variety of spaces (public and domestic, civic and ecclesiastical) and by a range of persons (lay and ordained, commoners and magistrates). For over a decade, the people of Augsburg were free to attend churches where the Eucharist was celebrated in different ways, each understood by its proponents as the one true and authentic way of doing so. Then, in 1537, the city council, calling on the authority of the Bible, abolished "the papist Mass," called for the removal of images associated with the mass, and dissolved the clerical estate. The council subsequently adopted its own "Church Ordinance," which prescribed a relatively simple "evangelical" order of service. In 1548, with Charles V's defeat of the Protestant alliance, Catholic worship was reimposed on Augsburg. Thus, for a time, people at many levels of society were free to deem different practices efficacious either by instituting new variations in practice or choosing the services they wished to attend. In 1537, the city council, backed, they argued, by the authority of the Bible, claimed the power to decide what practices were authentic. With the defeat of the alliance of Protestant cities and princes, the Catholic emperor overruled the city council's authority, and the Catholic mass was once again deemed efficacious (Wandel 2006, 46–93).

On the process whereby practices transcend time and place. The Augsburg case also demonstrates the complex processes that resulted in some practices coming to define translocal traditions, while others simply faded into obscurity. Thus, during the decade of relatively free experimentation, people enacted a variety of practices they deemed efficacious often without precise labels or formal written specifications. While the multitude of conceptions that emerged during this period could be characterized in retrospect as more or less Lutheran, Zwinglian, Catholic, Schwenkfeldian, or Anabaptist, they were characterized at the time simply as "Christian" or "evangelical" (48). The church ordinance promulgated by the city council in 1537 ended the period of experimentation and established a distinctive liturgy that fit none of these later designations precisely. This ordinance, though deemed efficacious by the council and enacted for eleven years, simply disappeared after 1548.

The practices that came to define translocal traditions—Lutheran, Reformed, and Catholic—were those that survived the vicissitudes of the wars of religion. In the wake of the wars, treaties established what practices were efficacious. Thus, the Augsburg Interim (1548) ordered Protestants to readopt Catholic practices, the Peace of Passau (1552) granted Protestants some freedoms, the Peace of Augsburg (1555) granted Lutheranism official status within the Holy Roman Empire, and the Peace of Westphalia (1648) did the same for Calvinism. Through this torturous process, Europeans decided that, within the bounds of the Holy Roman Empire, rulers could decide what practices would be deemed efficacious and thus what traditions could be established and defined within their own territories.

On the links between efficacy and experience. The way the Eucharist was understood had direct implications for the way that Christians were expected to experience Christ within each of the three traditions. The central divide between Catholics and Protestants had to do with whether or not Christ's sacrifice on the cross could be ritually duplicated in any way. Catholic teaching insisted that it could; Protestant teaching said that it could not. According to the Council of Trent, the Passover meal celebrated by Jesus and his disciples was to be interpreted in a broadly typological framework in which the "new" testament was understood to be prefigured in the "old." Thus, because of the "weakness of the Levitical priesthood," God ordained that "another priest should rise, according to the order of Melchisedech, our Lord Jesus Christ, who might consummate, and lead to what is perfect, as many as were to be sanctified" (the goal). "[T]hough he was about to offer Himself once on the altar of the cross [,] . . . His priesthood was not to be extinguished by His death." Thus, to effect the goal of sanctification, Jesus (as God), did the following:

> Declaring Himself constituted a priest for ever . . . , He offered up to God the Father His own body and blood under the species of bread and wine; and, under the symbols of those same things, He delivered (His own body and blood) to be received by His apostles, whom He then constituted priests of the New Testament; and by those words, Do this in commemoration of me, He commanded them and their successors in the priesthood, to offer (them); even as the Catholic Church has always understood and taught (Waterworth 1848, 153).

According to the Council, Jesus as priest offered up his own body and blood as a sacrifice at Passover, thus instituting the new Passover, reenacted by Jesus' successors in the priesthood. Thus, the same "divine sacrifice . . . is celebrated in the mass, [where] that same Christ is contained and immolated in an unbloody manner, who once offered Himself in a bloody manner on the altar" (154). The literal equation of the bread with Christ's body meant that then and in the present, "the victim is one and the same, the same now offering by the ministry of priests, who then offered Himself on the cross, the manner alone of offering being different" (155).

In this interpretation of "this do," the interpretation of "this" as a ritual sacrifice presided over by a priest is obviously crucial. The reenactment of the original sacrifice in the present is legitimated by a line of succession from Jesus to his disciples ("whom he constituted as priests of the New Testament") to priests properly ordained in a sacramentally constituted lineage. This lineage ensured that the priest, who celebrated the mass, was himself linked to Christ and could pronounce the words "this is my body" *in the role of Christ*, thus replicating not only the sacrifice but also the sacrificer sacrificing himself. It was not the priest as individual who effected the transformation of the bread and wine, but the words of Christ spoken by the priest in the role of Christ that did so. Thus, to sum up, in the Catholic view, an actual agent (the priest) must be ritually constituted (by the sacrament of ordination) so that he can take on the role of the ideal agent (Christ) whose word it is that transforms (Sørensen 2007, 85–87). The elements so transformed were understood "to contain the grace that they signify"—in this case, the power to sanctify—and to confer that grace effectively upon any who did not place "an obstacle" (Latin, *obex*: sinful act or disposition) in the way (Canon 7, Waterworth 1848, 55).

Luther and the Protestant reformers who followed him maintained a typological framework but configured it more narrowly, focusing on the words "this is the cup of a new testament in my blood which is shed for you for the forgiveness of sins" as the promise of a testator, rather than as the sacrifice of a priest. Thus, according to Luther:

[W]hat we call the mass is a promise of the forgiveness of sins made to us by God, and such a promise as has been confirmed by the death of the Son of God. . . . If the mass is a promise, as has been said, then access to it is gained, not with any works, or powers, or merits of one's own, but by faith alone. For where there is the Word of the promising God, there must necessarily be the faith of the accepting man" (Luther 1520, 2.45–46 [513–14], quoted in Wandel, 100).

Viewed in this light, Luther rejected the "conjoined" conception of the mass as a sacrifice offered by a priest, in which sacramental ordination conveyed the power to speak efficaciously the words that transformed the bread and wine into the body and blood. In its place he substituted the idea of the mass as a promise of forgiveness of sins that to be efficacious had to be received in faith by the communicant. Thus, according to Luther, "the mass is received by none but the person who believes for himself, and only in proportion to his faith" (Luther 1520, 2.68 [521–22]). In the Lutheran interpretation, the ritual subject (the communicant) must intentionally recognize (by faith) the power of the word to effect the transformation for the transformation (of the bread and wine) to be effective.

Though for both Luther and the Council of Trent it was the words of Christ that transformed the bread and wine, the shift in the proximate cause of the transformation from the priest to the communicant had a direct effect on the status of the materials transformed. While both Catholics and Lutherans believed that Christ was really present in the bread and wine, the shift in the (proximate cause) of that presence meant that Catholics understood Christ as present in all properly consecrated wafers while Lutherans understood Christ to be present only in consecrated wafers that were consumed by the faithful. With this shift, Luther rejected the whole panoply of Catholic paraliturgical sacramental devotions, and thus by extension a wide range of practices by means of which Catholics continued to expect that they could experience Christ as really present.

The central divide between Reformed Protestants, on the one hand, and both Catholics and Lutherans, on the other, concerned whether Christ could be seated "at the right hand of the Father" in heaven and at the same time be physically present in the eucharistic wafer. Catholics and Lutherans thought he could, though both traditions had difficulty explaining how he did it, while Reformed Protestants thought he could not. This meant that Catholics and Lutherans could expect to experience Christ as literally present in the eucharistic elements. The Reformed tradition, because it did not make a physical connection between the bread and the body of Christ, relied instead on the agency and power of the Holy Spirit to connect the believer with Christ's ascended body. Thus Calvin, and the Reformed traditions that followed him most closely, expected the Holy Spirit to carry believers "above all things that are visible, carnal and earthly," so that they could "feed upon the body and blood of Christ Jesus, once broken and shed for us but now in heaven and appearing for us in the presence of his Father" ("Scots Confession," quoted in Wandel, 188). In the Reformed traditions, following Calvin, it is the Holy Spirit that effects the communion between the believer and Christ's ascended body, through the faith of the communicant.

Different understandings of ritual efficacy, linked with different assumptions regarding how Christ could be present to the believer, created different normative expectations regarding experience. Catholics could expect to experience the real presence of Christ in the consecrated wafer, whether consumed during mass, displayed on the altar during the Benediction of the Blessed Sacrament, or carried in procession on the feast of Corpus Christi. Lutherans could only expect to experience the real presence when consuming the bread and wine in faith. Spiritually transported by the Holy Spirit, Reformed Protestants could expect to feed upon the body and blood of Christ in heaven. Thus, insofar as individual Catholics, Lutherans, or Reformed Protestants internalized the operative presuppositions of the traditions of which they were a part, those traditions would have led them to expect to experience Christ's presence in very different ways. If fully internalized, each tradition would have predisposed individuals to discount experiences of Christ's presence that did not fit with the normative expectations of their group and to assume the efficacy of their group's practices for effecting the real presence regardless of their particular feelings or perceptions on any given ritual occasion. What individual Catholics, Lutherans, or Reformed Protestants actually expected and/or experienced is, of course, another matter.

Reformation-era debates over the efficacy of practices led Protestants to formulate a distinction between religion and magic that allowed them to characterize Catholic practices as magical (improperly efficacious) and their own as religious. This distinction in turn heightened the Protestant distinction between ritual and experience and led Protestants to valorize experience along with scripture in determining what counted as authentically religious. Because these distinctions informed early social scientific definitions of religion and magic and still have a lingering hold on the study of religion, we need to consider the Reformation-era disputes over efficacy in light of both first-order (emic) and second-order (etic) definitions of magic. Several points can be clarified in light of the preceding analysis.

First, all parties viewed their own eucharistic practice as efficacious relative to the goal of communion with Christ. Although they offered different explanations of how their practices exerted their efficacy and set different requirements for achieving efficacy, they all ultimately attributed the efficacy of their own practices to divine agency, whether God, Christ, or the Holy Spirit.

Second, based on their disparate readings of the Bible, all parties concluded that their rivals' practices were not efficacious. Catholics and Protestants understood efficacy differently. For Catholics, efficacy was attributed to the words spoken by the priest insofar as the priest was understood to be speaking mimetically as Christ, a power granted to him by sacramental ordination. Sixteenth-century Protestants construed the mimetic efficacy accorded to the words of institution spoken by the priest as magical and contrasted it with the ostensibly nonmagical efficacy of faith in a promise contained in the words. In the context of the Reformation, Protestants hammered out a distinction between religion and magic in which magic was associated with practices deemed efficacious in themselves (that is, automatic) and religion with the (nonautomatic) power of the deity to effect what was promised when it was received in faith. This, of course, was a polemical distinction rooted in the Protestant rejection of the Catholic understanding of Jesus as priest (based on a typological reading of the Last Supper) and the mass as a sacrifice (for details, see the preceding analysis). This first-order definition of magic was in turn taken up in early social scientific definitions of religion and magic (Thomas 1971, 75–76, 81; Sørensen 2007, 9–10).

If we shift to a second-order definition of magic, such as that advanced by Sørensen (2007), we can reframe the Reformation-era debate differently. To summarize, Sørensen defines magical action as action that effects a change of state or essence in persons, objects, acts, or events through "certain special and non-trivial kinds of actions with opaque causal mediation" (32). Certain elements must be present to make the action efficacious, but, he argues, at least one element—an actor, an action, or an object—is invested with the "agency necessary for the ritual to have any effect" (65). He develops a typology of magical actions (transformative and manipulative) and frames of ritual action that link actions to specific goals (95–140). Though he does not provide a fully elaborated theory of the relationship between religion and magic, he suggests that magic, understood in this way, functions innovatively in relation to the elaborated symbolic interpretations of ritual advanced by religious traditions and thus is often a factor in the emergence of new religious movements. Magic from this perspective is not antithetical to religion but rather integral to it (181–91).[4]

---

[4] I concluded that adapting first-order terms such as "religious" or "magical" for second-order purposes generates more confusion than it resolves in response to reflecting on Sørensen's use of magic as a second-order concept and my own use of "religious" in an earlier draft (see the conclusion, below). Sørensen's very promising theory could easily be recast in

From this perspective, Reformation-era eucharistic debates were debates over what change was effected and what ritual element was necessary to bring about the effect. Catholics and Lutherans viewed objects (the bread and wine) as changing their state (becoming the body and blood of Christ) and thus becoming really present as such to properly receptive ritual subjects. Reformed Protestants viewed the ritual subject as somehow changing their state, such that they entered into communion with Christ in heaven (where he was "seated at the right hand of God"). Although all located ultimate agency in some aspect of the deity, Catholics located proximal magical agency in the sacramentally ordained priest, while Protestants shifted the locus of magical agency to the faith of the ritual subject. Viewed from the second-order perspective advanced by Sørensen, it was the magical efficacy that Protestants attributed to "faith" that fueled the emergence of new forms of Christianity (new religious traditions) with new ritual structures and symbolic elaborations.

This shift in the understanding of efficacy can also be interpreted as a shift in the locus of authority from the priest (institutional authority) to the lay priesthood of believers (individual authority) rooted in a more individualistic reading of Christian scripture. Though Luther quickly moved to moderate the individualistic effects of this shift, political and social dynamics made this difficult in the short term and impossible in the long term. The valorization of the individual in opposition to the group or its leadership is endemic to Protestantism. While individuals can claim authority on various grounds, scripture and experience are probably the two most common. If an individual takes a stand against the group on the grounds of experience, s/he is pitting individual experience deemed religious (a simple ascription) against the path deemed religious by the group (a composite ascription). The valorization of "religious experience" within the study of religion, particularly when understood individualistically, reinforced a Protestant bias and obscured the range of possible interactions between composite and simple ascriptions.

Simple and Composite

In comparisons of this sort, we can construct an object of study in which a feature common to both the simple and the composite formation forms the point of analogy. By comparing simple and composite formations, we can look at what people say they experience in relation to the formal

---

more generic terms. His "sacred" domain could be redescribed as a "special domain" and his theory of magic as a theory of ritual efficacy. Questions of efficacy involve causal explanations, both everyday and scientific, and thus can be framed in terms of attribution theory. Actions with "opaque causal mediation" are actions to which people attribute everyday explanations that scientists consider "opaque" or, in informal parlance, as "magical."

(normative) expectations of the path they are following. In particular we can look at what happens when people claim that they experience the originary event that the path seeks to re-create (the goal of the path) in a sensory or perceptual sense. Intracontextual comparisons of this sort allow us to focus on conflicts over the legitimacy of simple formations (what individuals claim they experience) relative to composite formations (what individuals claim they experience) relative to composite formations. Experiences that subjects consider religious may be embraced, channeled, or delegitimated by those positioned to speak for a tradition. Based on such experiences, subjects may launch movements of reform or renewal, some of which lead to the formation of new religious movements. The primary question for subjects is: Is this experience authentic—that is, real—and, following from that, who decides and on what grounds? As scholars, we can track their answers through an analysis of discourse and power relations.

Intercontextual comparisons of this sort allow us to compare the ways in which individuals and traditions negotiate the re-creation of originary events when individuals or groups report immediate sensory perceptions of the originary event. The questions for scholars include: How can we explain such experiences? Are there cross-culturally stable techniques that facilitate immediate sensory perceptions of originary events? How do individuals utilize these techniques, and how do authorities and other observers within traditions assess their effects? What effects do such experiential claims (simple formations) have on the tradition (composite formations)? What is the relation between the formal expectations of a path and sense perceptions of individuals?

In this case, a simple formation (visionary experiences) provides a point of analogy between two composite formations (medieval Catholicism and modern neopaganism). For the purposes of the comparison, I will specify the Catholic composite in terms of a sacramental path (practices deemed efficacious) relative to a special goal (salvation by an anomalous agent) with a particular focus, as in the previous section, on the Eucharist as a practice deemed efficacious for re-presenting Christ. I will specify the neopagan composite in terms of rituals (practices deemed efficacious) relative to a special goal (summoning powers on the magical plane, where the magical plane is understood as a different level of reality set apart by and operating according to its own rules, and thus protected from processes of secularization and disenchantment). In both cases practices of meditation and visualization form a link between the simple formation (visionary experiences) and the collective (ritual) practice that defines the composite.

In the Catholic case, practices of meditation and visualization elaborated in the monastic context were widely diffused among the laity in simplified form from the mid-twelfth century onward (Newman 2005). The dissemination of simplified meditation and visualization practices coincided with a changed understanding of the Eucharist consolidated at the Fourth Lateran Council (1215) with the promulgation of the doctrine of transubstantiation. In light of the new doctrine, Christ was understood to appear at the moment of consecration at the point when the priest said the words "Hoc est corpus meum," rather than at the moment of communion, as in earlier Catholic and later Protestant conceptions. This meant that whether or not believers "received" communion, they could expect to "meet Christ at the moment of his descent into the elements—a descent that paralleled and recapitulated the Incarnation" (Bynum 1987, 53). The resulting shift in emphasis from "receiving" Christ (communion) to "seeing" Christ in the host (adoration) was reflected in the rapid spread of new practices (the elevation of and acts of reverence toward the consecrated host) and new architectural spaces (Gothic cathedrals) during the thirteenth century. The new emphasis on seeing Christ in the host, when linked to practices of meditation and visualization, led to an outpouring of lay eucharistic visions in the late medieval period (Bynum 1987, 53–60; Camille 1996; Newman 2005, 16).

Among the most consequential eucharistic visions was that of Juliana of Cornillon (c. 1193–1258), a devout lay woman from the town of Liège in Belgium. She and her allies assiduously promoted her vision in support of their campaign to institutionalize a eucharistic feast honoring the consecrated host. The resulting Feast of Corpus Christi, approved by Pope Urban IV in 1264, spread widely throughout the church by the fourteenth century (Rubin 1991, 164–85).

Carolyn Bynum provides numerous examples of other eucharistic visions experienced by devout lay women, many of them beguines.[5] Hadewijch, a beguine from the Low Country, provided a first-person account of how Christ came to her "in the form and clothing of a man" at mass in the mid-1220s:

> Then he [as a man] gave himself to me in the shape of the sacrament [that is, as a consecrated host], in its outward form, as the custom is; and then he gave me to drink from the chalice. . . . After that he came himself to me, took me entirely in his arms, and pressed me to him; and all my members felt his in full felicity, in accordance with the desire of my heart and my humanity. So I was outwardly satisfied and fully transported. And then . . . after a short time . . . I saw him . . . so fade and all at once dissolve that I could no longer recognize or perceive him outside me, and I could no longer distinguish him within me. Then it was to me as if we were one without difference (quoted in Bynum 1991, 120).

[5] Beguines were uncloistered lay women following a devout life apart from an established rule, who lived in a variety of communal arrangements of varying degrees of formality (McGinn 1998, 32–41).

This vision, in which Christ himself offered her the consecrated bread and wine and then took her in his arms until they joined as one, was probably not a spontaneous vision, but rather the fruit of a complex spiritual discipline developed in medieval monasteries and brought to fruition by thirteenth-century beguines such as Hadewijch (Newman 2005, 3, 25).

Monasteries were an ideal context for generating visions. Monks and nuns were constantly immersed in scriptural and hagiographic texts replete with visions and engaged in practices—such as fasting, *lectio divina* (scriptural reading interspersed with meditation), the chanting of the Divine Office, and private prayer—conducive to visionary experiences. Monastic practices focused attention on sacred objects, such as a crucifix, an image, or the consecrated host, and encouraged the visualization of scriptural narratives. The practice of *lectio divina* combined the reading aloud of scripture with reflection on scripture (*meditatio*), mental or vocal prayer (*oratio*), and contemplation (*comtemplatio*). The process could be visually inflected, such that meditation involved visualizing the scriptural text and contemplation resulted in visions. These practices led to the proliferation of "unscripted" visions within monastic communities, many of which were recorded and shared (Newman 2005, 14–18).

Although many beguines adapted these techniques, they required a level of training unavailable to most of the laity. As the desire for more intense religious engagement spread in conjunction with the more visual emphasis in the mass, clerical writers began to produce texts to help lay readers visualize the life of Christ more vividly. These texts provided guided meditations—"visionary scripts"—that allowed ordinary people to visualize the life of Christ and then place themselves in the scenes they envisioned. These guided meditations easily crossed the line from visualizing to seeing, resulting in what Newman calls "scripted visions" (Newman 2005, 25–33).

There was a tension, however, between practices of visualization and meditation, which assumed a role for human agency in the promotion of divinely given visionary experiences, and the theology of the era, which viewed visions as the product of unmixed agency, whether divine, human, or diabolical. The tension between these two competing theologies, Newman argues, heightened late medieval clerics' anxieties about uncontrolled lay visionaries. As long as intentionally cultivated visions remained within the cloister, where they were construed as a form of piety and subordinated to the social constraints of monastic culture, they did not pose a threat. Cultivated outside the cloister, monastic piety was easily transformed into prophetic visions that laypersons could mobilize in opposition to authorities. With the spread of scripted visions among the laity in the later Middle Ages, clerical authorities came to view them with increased skepticism, attributing them more often than not to delusion (that is, human imagination) or diabolical agency (Newman 2005, 5–6, 33).

The criteria for distinguishing between divine, human, and demonic agency were not clear-cut. Even when visionaries downplayed the role of human agency in their narratives, there were no reliable criteria for discerning whether visions were divinely or diabolically inspired. By the fifteenth century, clerical writers intent on distinguishing between saints and witches found women's claims to divinely inspired visions increasingly suspect (Caciola 2003, 315–19; Klaniczay 2007). As Newman concludes:

> The position that finally triumphed . . . gave maximal authority to clerics charged with the discernment of spirits. By their standards the vast majority of reported visions would be judged inauthentic, while those that eventually passed muster as "private revelations" had to meet a daunting set of criteria designed to assure precisely that they had not been cultivated and did not stem from the visionary's imagination (41).

Human imagination and divine agency, which late medieval clerics were so eager to differentiate, are unabashedly conflated by neopagan traditions eager to align themselves with those the medieval church deemed heretical.

In the modern era, Jesuits and neopagans are two subcultures that retain a strong interest in practices of visualization and meditation. In an ethnographic study of modern witchcraft groups in suburban London in the early 1980s, Luhrmann (1989) found that "New Age" magicians relied on meditation and visualization practices as the foundation of their "craft." Gerald Gardner, a British civil servant, founded the groups that she studied in the 1940s. Gardner was likely inspired by Margaret Murray's claim that an organized pre-Christian cult of the goddess lay behind the European witchcraft persecutions as well as by turn-of-the-century occult groups, such as the Hermetic Order of the Golden Dawn (Luhrmann 1989, 42–43).

We can understand the ritual magic that constitutes the modern witch's "craft" as a special path. The goal of ritual magic is to summon powers that operate on the "magical plane." These powers, depicted in novels such as Tolkien's *Lord of the Rings* or Ursula LeGuin's *Earthsea Trilogy*, are understood as an amorphous elemental force that can be summoned by wizards, sorcerers, or magicians, though ultimately beyond their control. The primordial goddess is sometimes used to represent these powers in their creative, destructive, and transformative aspects (86–99) and ritual magic is deemed efficacious for summoning them.

These powers are believed to exist on a "magical plane" that is set apart from the everyday mundane world. According to Luhrmann, "the notion of a real-but-different magical reality is central for a modern magician, for it allows him both to assert the magic and to block it off from the skeptic's probing stare" (275). The association of the powers with a "magical plane" constituted by its own "reality" sets the powers apart and protects them from secular, rational critique.[6] Many of Luhrmann's informants viewed their magical practice as religious or spiritual (177), though not necessarily as "a religion" (337).

As in the late medieval period, modern magicians use both scripted and unscripted visualizations. In contrast to late medieval visualization practices, which were unscripted in the cloister and more likely scripted among the laity, Wiccan training moves from scripted to unscripted meditations. Magicians undergo a training process, often in the form of home-study courses, in which they complete a series of lessons involving daily exercises and short written assignments. In addition to explaining magical theory and practice, the lessons teach the student to meditate and visualize. Guided meditations, which the neophyte is instructed to memorize or record so they can listen to them in a relaxed state, lead the trainee into alternate worlds, where they meet and engage with guides and companions on the magical path. Luhrmann comments: "The striking aspect of the lesson[s] [in guided meditation] is the sense of reality which the authors clearly expect the exercitants to feel. And indeed with time the sense of reality does heighten." Based on her own experience of the lessons, she says, as "the depth of meditation increases, the imagery grows more intense" (182–83).

In contrast to the training exercises, the Wiccan rituals described by Luhrmann are unscripted and, in that sense, more like monastic than lay practices of meditation and visualization. She describes a ritual that took place at a retreat center over a long weekend in some detail. The ritual involved "working" with historical figures from Elizabethan England under the direction of an adept. The heart of the ritual was a guided meditation—a quasi-historical narrative—involving the sea voyage of the explorer William Drake and his crew aboard the *Golden Hind*. Although an individual adept narrated the meditation that formed the imaginative vehicle through which the power would be summoned, the general outline of the meditation was generated through a collective process of preparation. The group members, who had individually prepared for the ritual through structured meditations, as well as reading and research on Elizabethan topics, were in frequent contact regarding their dreams and meditations prior to the ritual. During this preparatory process they attributed intuitions and feelings that something was important to highly evolved beings that they viewed as guiding and directing their preparations (206–7).

---

[6] While Renaissance magicians had to defend themselves against charges of paganism and demon-worship, contemporary magicians seek to legitimate their practices in relation to scientific claims (Hanegraaff 2003, 369–71).

During the ritual, an adept narrated the meditation while participants sat in the dark with their eyes closed in order to focus their attention on the visualization and imagine it into reality (121).

Following the ritual, participants wrote up their experiences and sent them to the presiding adept, who compiled a general report containing excerpts from their accounts. The report, which drew selectively from the accounts, indicates: "The power came in very strongly, even when the hall was being prepared for this working. As one who assisted reports, 'during the setting out of the room, the ship began to materialize, almost, and with the most unexpected power. So much so that I found it difficult to walk, and found myself swaying with a rolling gait.'" Others quoted in the report also described sensations that they attributed to being at sea. Still others sensed figures (Drake's archetype, Arthur, the two Marys) or forces (Lemurian and Christian) (209). While not all personal accounts were quoted, Luhrmann doubts that the personal "reports lie, or consciously misrepresent the experience. The sense that there are 'contending forces' or a rocking ship was no doubt very 'real' and the experiential gist of these descriptions is no doubt accurate" (Luhrmann 1989, 209–10). Nonetheless, the inclusion of individual experiences that fit with the overall normative expectations of the group—that is, that what was visualized had become real—strengthened the group's confidence in its claims.

Both medieval Catholics and neopagan magicians used visualization practices, but the practices stood in different structural relation with their composite formation. In the Catholic composite the sacraments were, theoretically at least, the core practice deemed efficacious with respect to the goal of salvation. Among the sacraments, the Eucharist had pride of place. In relation to the composite, visualization and meditation were ancillary practices that the pious could use to enhance their engagement with the core practices. Even within the monasteries, where visualization and meditation practices were intensively cultivated, the liturgy of the hours was the core practice that, when added to the sacraments, defined monastic life. Within the Catholic Church, the only place where practices of meditation and visualization were integral to a (subsidiary) composite was in the Society of Jesus. Visualization practices were and are central to the spiritual exercises created by Ignatius of Loyola and given to his followers. Within the Jesuit composite the Ignatian spiritual exercises are the distinctive practice deemed efficacious relative to the goal of "making Jesuits."

The ancillary role that visualization and meditation practices play in the official Catholic composite since the Reformation stands in contrast to their role in neopagan groups. In the latter, meditation and visualization practices are deemed efficacious for summoning the powers that operate on the "magical plane." Church authorities could dissuade the Catholic laity from using meditation and visualization practices without altering the shape of the Catholic composite (though their suppression might dramatically alter the actual experience of the laity). The same cannot be said for neopagan magic. If visualization and meditation practices were suppressed, there would be no magic.

IMAGINATION AND REALITY

In the case of both medieval visionaries and neopagan magicians, we can see a cultivated movement from visualizing to seeing, which subjects experience as a movement from imagination to reality, from seeing "as if" one were present in the imagined world to a sense of actually being there (Newman 2005, 29). Subjects may view this sense of actually being there as real or imagined or, as was the case with the witches Luhrmann studied, as imagination turned real. The theme of imagination and reality also informed subjects' and researchers' interpretations of sleep paralysis and Reformation-era theologians' debates over the (real) presence of Christ in the Eucharist.

The interplay between imagination and reality has been a theme throughout this chapter and, indeed, throughout the book. The transition zone between what is imagined (as if) and what seems real (what is) is a point where the comparisons discussed in this chapter meet. They meet in the seeming reality of the felt presences and out-of-body experiences associated with sleep paralysis and the uncertainty, upon reflection, with regard to whether they really were. They meet in discussions of whether Christ is really present in the Eucharist and, if so, in what sense and by what means. They meet when cultivated imaginings (as if) turn to experienced realities (what is) in eucharistic devotions and magical rituals. This meeting, which often takes place in the context of ritual, has been the site of contestation both within traditions (emically) and among scholars (etically). Two problems confront us when we attempt to think about this interplay more deeply, difficulties inherent in any attempt to distinguish between imagination and reality and the intertwining of etic and emic perspectives. The discussion can move forward more effectively if we sort out the issues involved from the perspective of observers who position themselves either inside or outside a given composite formation (see table 4.2).

Emic versus etic observers. Emic observers are those who agree that a particular event should be deemed religious (thus constituting it as an originary event [OE]) and that a path can be constituted for re-creating the OE in the present. Though they may disagree over how the OE (the goal of the path) *should be* understood or what practices *should be* deemed efficacious in relation to the goal, they are nonetheless all *engaged* in defining

TABLE 4.2
Perspectives of Emic and Etic Observers on the Experience of Subjects

	Subject of Experience (Inside Composite)	Emic Observer (Positioned Inside Composite)	Etic Observer (Positioned Outside Composite)
Initial Situation	Subjects report immediate (on-line) sensory perception of OE (based on external or internal percepts) vs. not.	OE assumed; observer is positioned to distinguish b/w immediate sensory perception of OE that is congruent with normative expectations vs. not.	OE not assumed; observer is positioned to distinguish b/w an immediate sensory perception (that is, measurable in principle) of the alleged OE vs. no such experience
Analysis	Upon reflection subject may view exp. of OE as real (is) or imaginary (as if).	Authenticity of subject's experience must be determined.	Experience of subject and response of emic observer are taken as data to be analyzed.
Theoretical Aims	Subject engages as desired with perspectives of observers.	Theory explains whether experience of OE is real (authentic) or unreal (inauthentic).	Theory explains what makes experience of OE seem real to subject and response of emic observers.

the path and thus count as emic observers. Emic observers thus position themselves within the composite broadly defined by the OE, though they and/or others may view them as insiders or outsiders relative to particular formulations of paths related to an OE. Though Luther was excommunicated, and thus formally declared an outsider relative to the Catholic path, Lutheran, Catholic, and Reformed Christians, despite their disagreements, were all emically engaged in defining the Christian composite.

Etic observers are those who do not consider the event in question as special (that is, as an OE) and thus stand outside the composite broadly defined by the OE. If the originary event is understood as deeming Jesus as the Messiah, then Jews and Muslims (along with atheists, Buddhists, and many others) stand outside the composite defined by that OE. Where one stands in relation to any given composite can be understood as a matter of position or perspective rather than essence. Observers who view their stance as position may observe a subject *as if* they deemed an event religious (emically) or *as if* they did not (etically). Scholars of religion of various religious persuasions or none at all frequently move between these two stances, cultivating the ability to examine a composite as if from within (emically) and as if without (etically) depending on their relation to it.

Judging authenticity. Officially and unofficially, persons associated with traditions make claims about the nature of reality and how it can be known. Emic observers judge individuals' claims to have had real (authentic) sensory perceptions of originary events in light of those criteria. Within the tradition, authorities and individuals may disagree over these determinations and authorities may disagree with one another. They may declare the experiences in question false or delusional or mere products of the imagination, rather than as authentic experiences of the originary event. Only emic observers are capable of making determinations of authenticity. This is simply a matter of logic, not policy. Etic observers, because they do not view the events in question as originary, simply have no criteria for judging whether a sensory perception authentically reproduces an originary event or not. Although observers cannot argue for or against the authenticity of a re-creation of an OE from an etic perspective, etic observers can and frequently do argue that a claim is delusional—that is, an incorrect inference about external reality—on the grounds that an event (taken specifically or generally) should not be deemed religious, and thus that no practice is capable of re-creating it.

Analyzing experiences. Based on the sorts of evidence for accessing experience discussed in chapter 2, both emic and etic observers of a tradition can explore a series of empirical questions that stop short of determinations of authenticity, though they will usually do so with different motivations and interests. Both can attempt to determine whether or not an individual actually had an immediate sensory perception of what the subject took to be the OE. Observers can attempt to establish whether or not the subject experienced immediate sensations or perceptions that they associated with the originary event. In principle, researchers could measure these sensations or perceptions and/or establish their neural correlates. Such experiences can be distinguished in principle from experiences in which subjects claim to engage the originary event in the absence of immediate sensations or perceptions (for example, "through faith" or "as if" the OE were re-created in the present). We can distinguish, in other words, between someone like Hadewijch, to whom Jesus appeared as a man in the context of the mass, and someone who devoutly genuflects at the elevation of the host but sees nothing unusual. Although both etic and emic observers could pursue this distinction, emic observers might be more interested in what the subject took to be the OE (and whether it conformed to normative expectations) than in how the subject experienced it. Both might be interested in knowing whether subjects, upon reflection, viewed their sensory experience of the OE as real or imagined.

Although both insiders and outsiders are able to think in sophisticated ways about the boundary between the real and the imaginary, they typically do so in different ways given their differing assumptions regarding the OE. Insiders typically turn to matters of authenticity—that is, to criteria for discerning the authenticity of the experience in light of the beliefs that they hold with respect to the OE—while scholars taking an etic stance typically try to explain what made the experience of the OE *seem real to the subject.* We can use these distinctions in conjunction with the levels of analysis discussed in chapter 3 to identify three different contexts in which issues of imagination and reality come to the fore: in relation to collective rituals; in relation to specific practices (for example, visualizations and meditations); and in relation to seemingly spontaneous experiences (for example, sleep paralysis). In each case, there are relevant bodies of research that can illuminate the subject's experience of the interplay between imagination and reality from an etic perspective.

How might experience seemingly become real to subjects in the context of ritual? Luhrmann (1989) and Sharf (2005) both reflect on this question in relation to Wiccan (Luhrmann) and Buddhist (Sharf) ritual and arrive at similar conclusions. Both encourage us to think of ritual as a special form of adult play. Play theorists describe play as an activity in which one manipulates latent metalinguistic cues in order to construct an *as if* world. The world of play can be bounded by rules or simply by the sense of "as if" or "let's pretend." Both ground adult ritual play developmentally in the work of psychologists, such as Vygotsky (1978) and Winnicott (1971), who view play as constituting an ambiguous intermediate space between the subjectivity of the self and the objectivity of the other, the space in which an inherently social sense of self emerges (Luhrmann 1989, 331–36; Sharf 2005, 253–57). Hanegraaff (2003), building on suggestions offered by Luhrmann (1989) and classical anthropological theorists, characterizes what goes on in play as rooted in a spontaneous tendency of mind, which he refers to as "participation" and contrasts with "instrumentality." He conceives of the one as a kind of "spontaneous animism" and the other as a spontaneous skepticism, an opposing tendency to explain things in terms of material causation (Hanegraaff 2003, 371–76). In this view, ritual as play creates a privileged space in which the former can be cultivated.

How might experience seemingly become real to subjects by means of specific techniques? Meditation, which focuses attention, and visualization, which draws people into an image or scene, can both be used to enhance mental imagery and, by extension, the reality of the play world. Though some individuals may have more natural ability to create vivid mental images, this ability can be cultivated through the use of these techniques. The use of techniques to cultivate mental imagery is found in most cultures (Noll 1985). Recent research on "false memories" of childhood sexual abuse, ritual abuse, and alien abduction suggests that hypnosis is a related technique that can also be used to generate vivid mental images. The therapeutic relationship can be construed as another interactive "play space." In that context, hypnosis sometimes leads to the creation of "memories" that seem to be "recovered" rather than imagined (Schacter 1995; De Rivera and Sarbin 1998; Clancy 2005). In a case of alleged satanic ritual abuse that came to trial in the early 1990s, the accused man "remembered" what had allegedly occurred only after learning how to meditate and pray about each of the new accusations. A psychologist called in by the investigators made up an accusation that he told the accused had been reported to him by the accusers and, after praying about it, the accused produced a "memory" of the fictitious incident (Wright 1994, 134–46).

How might experience seemingly become real to subjects spontaneously? Dreams provide an instructive context for considering this question. Dream content, like sleep paralysis, is disproportionately weighted toward themes of threat and predation. Revonsuo (2000) theorizes that the biological function of dreaming is to simulate threatening events and to rehearse threat-perception and threat-avoidance. Cheyne (2000), in response to Revonsuo, argues that the substantial minority of dreams where themes of threat and predation are absent suggests a broader hypothesis that links dreams to play. He theorizes that dreams allow individuals to simulate threats and other unusual situations in which practice can improve the individual's ability to respond. Dreams thus may provide a safe space in which to test the limits of our ability to respond under exaggerated and unusual conditions. Our test case— Barnard's sense of being lifted outside himself—was triggered by the attempt to visualize (simulate) an extreme situation, specifically his attempt to envision his self-awareness not existing after his death. Though his experience was triggered by informal visualization and took place during the day, it would seem likely that it involved processes of this sort.

Religions

A BUILDING-BLOCK APPROACH

I had two insights in the course of writing this book that fundamentally altered my sense of how we ought to study religion. The first arose as I began to pay attention to the way scholars use terms related to religion. While we routinely refer to the study of religion and definitions of religion, I noticed that in switching to an ascriptive formulation, I was forced to use the adjective "religious" rather than the noun "religion." Moreover, in drawing on Durkheim's definition of "the sacred" as things set apart and prohibited, I realized that he used his definition of "the sacred" to define *a religion* rather than *religion* per se. In fact, he makes a very clear distinction between "sacred things" and "religions." In his words, it is only "when a certain number of sacred things have relations of coordination and subordination with one another, so as to form a system that has a certain coherence and does not belong to any other system of the same sort, [that] . . . the beliefs and rites, taken together, constitute a religion" (Durkheim 1912/1995, 38). He then adds: "By this definition, *a religion* is not necessarily contained within a single idea and does not derive from a single principle that may vary with the circumstances it deals with, while remaining basically the same everywhere. *Instead, it is a whole formed of separate and relatively distinct parts*" (38, emphasis added). Durkheim's distinction between "sacred things" and "religions" and his conception of religions as wholes formed of separate and relatively distinct parts led to the distinction between simple and composite ascriptions advanced in chapter 1.

The second insight, which came much later, had to do with the idea of "specialness" and its potential benefits as a second-order concept. This insight arose as I reflected on Sørensen's (2007) use of "magic" and "sacred" as second-order terms in relation to my use of "religious" as a second-order term in the first draft of this book. As I struggled to figure out how to discuss his theory of ritual, which he describes etically as a theory of "magic" and which he and I both think offers a coherent theory of how a great deal of "religious" ritual works, I finally decided it was just too confusing to use terms such as "magic," "sacred," and "religious" as second-order terms. In doing so, we wind up having to translate between second-order, scholarly discourses, when what we need is a common, more generic discourse that will allow us to analyze how people use these various terms in practice. In struggling with his definition of the "sacred domain," however, I was struck by his repeated use of the term "special."[1] Although it obviously encompasses much more than what people mean by "religious," "sacred," "magical," et cetera, I consider this as one of its virtues; it identifies a large conceptual domain in which religion-like concepts play an important but not definitive role. As such, it has the potential to connect the study of religion with (rather than isolating it from) other disciplines in the humanities and sciences.

The two insights, taken together, have left me with a much more expansive view of the way I think we ought to approach the study of religion. Whether—to borrow from Sørensen—we are talking about "*special* beings violating ordinary ontological assumptions, *special* and privileged discursive repertoires, [or] *special* modes of interaction" (2007, 63; emphasis added), special things of one sort or another are the building blocks of what we think of as religions or spiritualities. Religions and spiritualities are composite formations or, in Durkheim's terms, wholes made up of separate and relatively distinct parts. Focusing our attention on "special things" takes our attention away from "religion" in the abstract and refocuses it on the component parts or building blocks that can be assembled in various ways to create more complex socio-cultural formations, some of which people characterize as "religions" or "spiritualities" or "paths." A building block approach to religions, grounded in the concept of specialness and processes of singularization, strikes me as a more promising way forward in the study of religion than continuing to wrestle with defining the abstract concept of "religion."

Building Blocks

Things that strike people as special are (among) the basic building blocks of religion. Ascriptions of specialness may take place below the threshold of awareness; when this happens, it tends to make things seem inherently special. People can decide, upon reflection, that things that seem special are more or less special than they initially seemed. In the process of reflection, special things may be caught up in preexisting systems of belief and practice, may generate new or modified beliefs and practices, or may lose their specialness and become ordinary. Whether people consider a special thing as (say) "religious," "mystical," "magical," "superstitious,"

---

[1] Sørensen defines the sacred domain as a "conceptual domain" to which participants grant "*special*" status. He sums up his discussion of the sacred domain by saying that "the sacred domain involves *special* beings violating ordinary ontological assumptions, *special* and privileged discursive repertoires, and *special* modes of interaction" (2007, 63; emphasis added).

"spiritual," "ideological," or "secular" will depend on the preexisting systems of belief and practice, the web of concepts related to specialness, and the way that people position themselves in a given context.

If specialness strikes others as a plausible generic framework within which to consider culture-specific webs of religion-like concepts, this idea could be tested and further refined historically and cross-culturally. I have identified several marks of specialness—that is, prohibitions against trading, mixing, and comparing—and two types of things likely to be considered special, ideal things and anomalous things. I hope that scholars would continue to refine these marks and types and add others as needed to reflect what they find in various contexts. In the process, we should try to distinguish between culturally specific distinctions and underlying types of distinctions that may reflect basic affective and cognitive processes common to humans and perhaps to other animals as well. Scientists, borrowing a culinary metaphor, speak of carving various domains (for example, linguistics, biology) at their joints, by which they mean making conceptual distinctions that reflect underlying evolutionary and developmental processes and which are, in that sense, non-arbitrary. By comparing the way humans across time and cultures identify and mark things as special and by extending our comparisons, where relevant, to other species, we may find ways to carve specialness at its joints. In doing so, we would dramatically refine our understanding of one of the building blocks commonly used to create what we think of as religions or spiritualities.

The nature and extent of the role that unusual experiences have played in basic ascriptions and especially in the basic ascriptions that are most likely to be incorporated in composite formations remains an open question. In my view, some researchers have been too quick to identify unusual experiences as the primary source of religion (for example, experiences associated with sleep paralysis, trance, or hallucinogenic drugs). Clearly, however, unusual experiences of various sorts have played a large role in initial ascriptive claims. Moreover, many of these unusual experiences fall in the gray area between sleep and waking, either literally or conceptually. Distinguishing between simple ascriptions and composite formations allows us to develop our understanding of the role of unusual experiences in basic religious ascriptions in tandem with advances in cognitive and affective neuroscience. Doing so may allow us to develop a cognitively and/or affectively based typology of experiences often deemed religious.

Such a typology, if it can be constructed, will probably not look like classic typologies, but will have to reflect the fluidity associated with both discursive representations and brain processes. Indeed, envisioning a new approach to a cognitively based typology of experiences commonly deemed religious may be a point where the insights of critical theory and cognitive science converge. Discursively, we know not only that experiences deemed religious can morph into experiences that are not deemed religious and vice versa, but that the basic shape of the experience can change over time as well (Taves 1999, 2009b). This is at least as true at the level of brain functioning, where researchers stress that states of mind, including emotions and experiences, do not have discrete boundaries and that attempts to represent them linguistically are always approximate and context-dependent (Azari 2004, 47; Malle 2005b, 24–25). Still, experience morphs within constraints and takes recurrent forms. A new sort of typology might need to combine phenomenological descriptions of experiences that are relatively stable across cultures with a dynamic model of brain processes, such as Hobson's AIM model (see fig. 2.2; Hobson et al. 2000a, 836–41; Hobson 2001, 45–48, 133–52), which would allow us to map experiences in relation to intersecting physiological continua.

As new technologies make it possible to develop a comparative neurophysiology of altered states of consciousness, careful depiction of subjects' corresponding conscious experience, motivation, and intentionality will be crucial (Hobson 2001, 104). Additionally, as these typologies are refined, we could identify existing scales or questionnaires that would allow us to determine the prevalence of such experiences in the general population and the frequency with which religious ascriptions are applied to them. Existing scales or surveys may not be adequate for these purposes and new ones may need to be developed.

Religions as Composite Formations

Religions, spiritualities, and other paths can be understood as composite formations premised on a set of two or more interlocking ascriptions, at least one of which is a basic ascription with religion-like qualities. I have used the idea of a "special path" to illustrate the idea of a composite. A "path" is a shorthand way to refer to practices deemed efficacious relative to a goal. In this formulation rituals are a subset of practices, since there are practices that we might not consider rituals—for example, searching—that may be deemed efficacious relative to a goal such as "finding religious significance." We can also identify a subset of practices, or perhaps more precisely techniques, that manipulate the mind more directly than others. We examined techniques for manipulating attention (meditation) and sight (visualization) in some detail. It should be possible to develop a more elaborate, cognitively based typology of such techniques. Such a typology would undoubtedly also include techniques for manipulating sound (chanting, singing, recitation) and consciousness (hallucinogenic drugs). These consciousness-altering practices can relate in different ways to a composite. In some cases, consciousness-altering techniques are an integral part of the practices deemed efficacious relative to the goal: for example, peyote in relation to the Native American Church or meditation in relation to Americanized Zen practice. In other cases, such techniques inform ancillary practices that may be thought to enhance the core practices. Meditation practices played this role in relation to the Catholic sacramental system.

In this book I used special paths to illustrate the idea of a composite. I am assuming that this formulation of a composite could be much more fully developed and that other useful composites can be identified. Although such development is beyond the scope of this book, I offer a few preliminary thoughts in this regard in appendix C.

IMPLICATIONS

This building block approach has several implications for scholars of religion in departments of religion or religious studies.

1. I do not think we need to worry so much about defining "religion." I think we can simply consider it as an abstraction that many use to allude to webs of overlapping concepts that vary from language to language and culture to culture. I have suggested that the concepts that we (as scholars) associate with these webs are associated with things that people view as special and often, but not always, connect with special beliefs and special practices. Conceiving of "religion" in this way allows scholars to examine what things people consider special and how they position them in relation to these webs of concepts.

2. I do not think scholars of religion have a monopoly on special things, since there are lots of special things that do not have religion-like connotations, but I think it is quite possible that the more special people consider something to be the more likely they and others are to place it under some religion-like heading (for example, "religious," "sacred," "magical," "superstitious," et cetera). If that is the case, departments of religious studies might want to conceive of themselves as loci for studying special things and the ways people incorporate them into their lives and perpetuate them by means of larger socio-cultural formations (for example, religious traditions, spiritual disciplines, and other assorted paths). An underlying focus on specialness and processes of singularization would not provide religious studies with an exclusive franchise, but rather with a focus on processes that are integral to and exemplified in the formation of religions and spiritualities and, at the same time, extend well beyond them. A focus on such processes would provide a bridge to other disciplines across the humanities and the sciences.

Appendixes

General Attribution Theory of Religion

(Spilka, Shaver, and Kirkpatrick 1985)

A.1. People seek to explain experiences and events by attributing them to causes—that is, by "making causal attributions."

C.1.1. Often, an event or experience has many possible and perhaps compatible causes, in which case the attributor's task is to choose among them or rank them in terms of relative importance or causal impact.

C.1.2. In cases where the presumed causal agent is a human or humanlike actor, attributions are frequently made to some enduring trait(s) or other characteristic(s) of the actor.

C.1.3. In cases where the presumed causal agent is an actor, attributions are frequently made to the actor's reason(s) or intention(s).

A.2. The attribution process is motivated by (1) a need or desire to perceive events in the world as meaningful, (2) a need or desire to predict or control events, and (3) a need or desire to protect, maintain, and enhance one's self-concept and self-esteem.

C.2.1. Attributional activity consists in part of an individual's attempt to understand events and interpret them in terms of some broad meaning-belief system.

C.2.2. Attributional activity consists in part of an individual's attempt to maintain effective control over events and experiences, in order to increase the probability of positive outcomes and avoid negative outcomes.

C.2.3. Attributional activity consists in part of an individual's attempt to maintain personal security and a positive self-concept, including a general striving toward self-enhancement and the protection of both the physical self and the self-concept against threat.

A.3. Attributional processes are initiated when events occur that (1) cannot be readily assimilated into the individual's meaning-belief system, (2) have implications regarding the controllability of future outcomes, or (3) significantly alter self-esteem either positively or negatively.

A.4. Once the attribution process has been engaged, the particular attributions chosen will be those that best (1) restore cognitive coherence to the attributor's meaning-belief system, (2) establish a sense of confidence that future outcomes will be satisfactory or controllable, and (3) minimize threats to self-esteem and maximize the capacity for self-enhancement.

A.5. The degree to which a potential attribution will be perceived as satisfactory (and hence likely to be chosen) will vary as a function of (1) characteristics of the attributor, (2) the context in which the attribution is made, (3) characteristics of the event being explained, and (4) the context of the event being explained.

A.6. Systems of religious concepts offer individuals a variety of meaning-enhancing explanations of events—in terms of God, sin, salvation, and so forth—as well as a range of possibilities for enhancing feelings of control and self-esteem (for example, personal faith, prayer, rituals, and so forth).

C.6.1. Systems of religious concepts provide individuals with a comprehensive, integrated meaning-belief system that is well adapted to accommodate and explain events in the world.

C.6.2. Systems of religious concepts satisfy the individual's need or desire to predict and control events, either through mechanisms for directly influencing future outcomes (extrinsic form) or through suspension or relinquishing of the need for direct control (intrinsic form).

C.6.3. Systems of religious concepts possess a variety of means for the maintenance and enhancement of self-esteem, including unconditional positive regard, conditional positive regard, and opportunities for spiritual growth and development.

D.1. The likelihood of choosing a religious rather than a nonreligious attribution for a particular experience or event is determined in part by dispositional characteristics of the attributor such as (1) the relative availability to that person of religious and naturalistic meaning-belief systems, (2) beliefs about the relative efficacy of religious and naturalistic mechanisms for controlling events, and (3) the relative importance of religious and naturalistic sources of self-esteem.

D.2. The attributor's context influences the relative likelihood of a religious rather than a nonreligious attribution by temporarily altering (1) the relative availability of the two meaning-belief systems, (2) the attributor's perception of the efficacy of religious versus naturalistic control mechanisms, or (3) the relative salience of these competing sources of self-esteem.

D.3. Characteristics of events that influence the choice between religious and nonreligious attributions include (1) the degree to which the event to be explained is congruent with the individual's meaning-belief systems, (2) the degree to which religious and naturalistic mechanisms for controlling similar events are seen to be effective, and (3) the degree to which religious and naturalistic explanations represent potential sources of self-esteem.

D.4. The context in which a to-be-explained event occurs influences the likelihood of religious versus naturalistic attributions by (1) affecting the relative plausibility or availability of various explanations, (2) providing information about the efficacy of various mechanisms for potentially controlling similar events, and/or (3) influencing the degree to which the events impacts on the attributor's self-esteem.

Personal Accounts of Stephen Bradley and William Barnard

(James 1902/1985, 157–60; Barnard 1997, 126–29)

William Barnard: "When I was thirteen years old, I was walking to school in Gainesville, Florida, and without any apparent reason, I became obsessed with the idea of what would happen to me after my death. Throughout that day I attempted to visualize myself as not existing. I simply could not comprehend that my self-awareness would not exist in some form or another after my death. I kept trying, without success, to envision a simple blank nothingness. Later, I was returning home from school, walking on the hot pavement next to a stand of pine trees less than a block from my home, still brooding about what it would be like to die. Suddenly, without warning, something shifted inside. I felt lifted outside of myself, as if I had been expanded beyond my previous sense of self. In that exhilarating and yet deeply peaceful moment, I felt as if I had been shaken awake. In a single, 'timeless' gestalt, I had a direct and powerful experience that I was not just that young teenage boy but, rather, that I was a surging, ecstatic, boundless state of consciousness" (127–28).

Stephen Bradley: "Before entering upon a minuter study of the process, let me enliven our understanding of the definition by a concrete example. I choose the quaint case of an unlettered man, Stephen H. Bradley, whose experience is related in a scarce American pamphlet.[1]

I select this case because it shows how in these inner alterations one may find one unsuspected depth below another, as if the possibilities of character lay disposed in a series of layers or shells, of whose existence we have no premonitory knowledge.

Bradley thought that he had been already fully converted at the age of fourteen.

---

[1] James cites the source of the account as: A sketch of the life of Stephen H. Bradley, from the age of five to twenty four years, including his remarkable experience of the power of the Holy Spirit on the second evening of November, 1829. Madison, Connecticut, 1830.

I thought I saw the Saviour, by faith, in human shape, for about one second in the room, with arms extended, appearing to say to me, Come. The next day I rejoiced with trembling; soon after, my happiness was so great that I said that I wanted to die; this world had no place in my affections, as I knew of, and every day appeared as solemn to me as the Sabbath. I had an ardent desire that all mankind might feel as I did; I wanted to have them all love God supremely. Previous to this time I was very selfish and self-righteous; but now I desired the welfare of all mankind, and could with a feeling heart forgive my worst enemies, and I felt as if I should be willing to bear the scoffs and sneers of any person, and suffer anything for His sake, if I could be the means in the hands of God, of the conversion of one soul.

Nine years later, in 1829, Mr. Bradley heard of a revival of religion that had begun in his neighborhood. 'Many of the young converts,' he says,

would come to me when in meeting and ask me if I had religion, and my reply generally was, I hope I have. This did not appear to satisfy them; they said they KNEW THEY had it. I requested them to pray for me, thinking with myself, that if I had not got religion now, after so long a time professing to be a Christian, that it was time I had, and hoped their prayers would be answered in my behalf.

One Sabbath, I went to hear the Methodist at the Academy. He spoke of the ushering in of the day of general judgment; and he set it forth in such a solemn and terrible manner as I never heard before. The scene of that day appeared to be taking place, and so awakened were all the powers of my mind that, like Felix, I trembled involuntarily on the bench where I was sitting, though I felt nothing at heart. The next evening I went to hear him again. He took his text from Revelation: "And I saw the dead, small and great, stand before God." And he represented the terrors of that day in such a manner that it appeared as if it would melt the heart of stone. When he finished his discourse, an old gentleman turned to me and said "This is what I call preaching." I thought the same, but my feelings were still unmoved by what he said, and I did not enjoy religion, but I believe he did.

I will now relate my experience of the power of the Holy Spirit which took place on the same night. Had any person told me previous to this that I could have experienced the power of the Holy Spirit in the manner which I did, I could not have believed it, and should have thought the person deluded that told me so. I went directly home after the meeting, and when I got home I wondered what made me feel so stupid. I retired to rest soon after I got home, and felt indifferent to the things of religion until I began to be exercised by the Holy Spirit, which began in about five minutes after, in the following manner:—

At first, I began to feel my heart beat very quick all on a sudden, which made me at first think that perhaps something is going to ail me, though I was not alarmed, for I felt no pain. My heart increased in its beating, which soon convinced me that it was the Holy Spirit from the effect it had on me. I began to feel exceedingly happy and humble, and such a sense of unworthiness as I never felt before. I could not very well help speaking out, which I did, and said, Lord, I do not deserve this happiness, or words to that effect, while there was a stream (resembling air in feeling) came into my mouth and heart in a more sensible manner than that of drinking anything, which continued, as near as I could judge, five minutes or more, which appeared to be the cause of such a palpitation of my heart. It took complete possession of my soul, and I am certain that I desired the Lord, while in the midst of it, not to give me any more happiness, for it seemed as if I could not contain what I had got. My heart seemed as if it would burst, but it did not stop until I felt as if I was unutterably full of the love and grace of God. In the mean time while thus exercised, a thought arose in my mind, what can it mean? and all at once, as if to answer it, my memory became exceedingly clear, and it appeared to me just as if the New Testament was placed open before me, eighth chapter of Romans, and as light as if some candle lighted was held for me to read the 26th and 27th verses of that chapter, and I read these words: "The Spirit helpeth our infirmities with groanings which cannot be uttered." And all the time that my heart was a-beating, it made me groan like a person in distress, which was not very easy to stop, though I was in no pain at all, and my brother being in bed in another room came and opened the door, and asked me if I had got the toothache. I told him no, and that he might get to sleep. I tried to stop. I felt unwilling to go to sleep myself, I was so happy, fearing I should lose it—thinking within myself

"My willing soul would stay
In such a frame as this."

And while I lay reflecting, after my heart stopped beating, feeling as if my soul was full of the Holy Spirit, I thought that perhaps there might be angels hovering round my bed. I felt just as if I wanted to converse with them, and finally I spoke, saying "O ye affectionate angels! how is it that ye can take so much interest in our welfare, and we take so little interest in our own." After this, with difficulty I got to sleep; and when I awoke in the morning my first thoughts were: What has become of my happiness? and, feeling a degree of it in my heart, I asked for more, which was given to me as quick as thought. I then got up to dress myself, and found to my surprise that I could but just stand. It appeared to me as if it was a little heaven upon earth. My soul felt as completely raised above the fears of death as of going to sleep; and like a bird in a cage, I had a desire, if it was the will of God, to get released from my body and to dwell with Christ, though willing to live to do good to others, and to warn sinners to repent. I went downstairs feeling as solemn as if I had lost all my friends, and thinking with myself, that I would not let my parents know it until I had first looked into the Testament. I went directly to the shelf and looked into it, at the eighth chapter of Romans, and every verse seemed to almost speak and to confirm it to be truly the Word of God, and as if my feelings corresponded with the meaning of the word. I then told my parents of it, and told them that I thought that they must see that when I spoke, that it was not my own voice, for it appeared so to me. My speech seemed entirely under the control of the Spirit within me; I do not mean that the words which I spoke were not my own, for they were. I thought that I was influenced similar to the Apostles on the day of Pentecost (with the exception of having power to give it to others, and doing what they did). After breakfast I went round to converse with my neighbors on religion, which I could not have been hired to have done before this, and at their request I prayed with them, though I had never prayed in public before.

I now feel as if I had discharged my duty by telling the truth, and hope by the blessing of God, it may do some good to all who shall read it. He has fulfilled his promise in sending the Holy Spirit down into our hearts, or mine at least, and I now defy all the Deists and Atheists in the world to shake my faith in Christ.

So much for Mr. Bradley and his conversion, of the effect of which upon his later life we gain no information. Now for a minuter survey of the constituent elements of the conversion process . . ."

Preliminary Thoughts on the Elaboration of Composite Formations

In order to see if I could identify other composite formations, I analyzed a few classic and contemporary definitions of religions. What I learned from this exercise is that definitions have to be constructed ascriptively in order to work in a building block approach. Functionalist definitions do not work because their overarching structure is not ascriptive. The following three definitions illustrate the problem.

Durkheim.   Durkheim defines a religion as "a unified system of beliefs and practices relative to sacred things, that is to say, things set apart and forbidden—beliefs and practices which unite into one single moral community called a Church, all those who adhere to them." We can paraphrase Durkheim as follows: *A religion is a unified system of beliefs and practices relative to special things, which beliefs and practices unite into one single moral community [the goal] all those who adhere to them.*

In the special-path composite, the goal of the path relates to something special. In Durkheim's formulation, the "system" replaces the "path" and it is the "system" that relates to special things, not the goal. Uniting as a moral community is not the conscious goal of those who adhere to the system of beliefs and practices and therefore the system of beliefs and practices is not explicitly deemed efficacious relative to the goal by adherents. Adherents presumably adhere to the system because it relates to special things. This shift, in which the scholar rather than the adherents identifies the goal, may be the key to constructing a passive functionalist composite in which something (the system) effects a goal (unifying into community) that adherents are not consciously pursuing.

Geertz.   We see something of the same situation with Geertz's definition. Geertz (1973, 4) defines a religion as "(1) a system of symbols which acts to (2) establish powerful, pervasive, and long-lasting moods and motivations in men by (3) formulating conceptions of a general order of existence and (4) clothing these conceptions with such an aura of factuality that (5) the moods and motivations seem uniquely realistic." We can rearrange Geertz's definition to read: *A religion is a system of symbols that relates to a conception of reality (a general [transcendent] order of existence clothed with an aura of factuality), which acts to establish dispositions that seem uniquely realistic [to adherents].* Here we have another functionalist definition in which something Geertz deems special (a transcendent conception of reality) effects a goal (establishing dispositions that seem realistic to adherents) that the adherents are not consciously pursuing.

Tweed. Tweed (2006, 54) provides a more recent functionalist definition that focuses on space and movement. "Religions are confluences of organic-cultural flows that intensify joy and confront suffering by drawing on human and suprahuman forces to make homes and cross boundaries." We can paraphrase Tweed's definition to read:

Religions are flows that intensify joy and confront suffering by drawing on special forces (means) to make homes and cross boundaries (goal). In this definition, people draw on special forces in order to pursue the goals of finding a place and crossing between things. If the people do not consciously pursue the goals of dwelling and crossing, then this is another functionalist definition. If people consciously draw on special forces to find special places and cross over special boundaries, then we have a composite in which people draw upon special forces (means considered efficacious) relative to a goal.

Although they may suggest components that could be incorporated into a workable composite, functionalist definitions of religions cannot be employed as formulated in a building block approach to religions because they focus on how the scholar thinks religions *function* (to unite [Durkheim], to establish dispositions [Geertz], to orient in time and space [Tweed]) rather than on how people put religions together. In contrast, although Weber did not formally define religions, he discusses them as "paths of salvation." His formulation reflects and elaborates on the *marga* form. As such, it can easily be used in a building block approach to the study of religions.

Weber. In the "paths-of-salvation" formulation, salvation designates the goal as one of redeeming adherents from something and for something, where "from what" and "for what" depend upon one's image of the world (Gerth and Mills 1958, 280–81). Those on the path always explicitly seek the goal and by implication deem the path efficacious with respect to the goal. Weber's composite can be paraphrased to read: *Religions are paths deemed efficacious relative to the goal of redeeming adherents from something and for something.* Whereas the special-paths formulation used in this book specifies the path in terms of practices deemed efficacious, the paths-of-salvation formulation focuses on the transformative character of the goal.

Weber provides a long list of possible instantiations of goals that redeem from something for something, all of which fall under the heading of "special things" (figure 1.1). These features are highlighted below in italics.

- One could wish to be saved from political and social servitude and lifted into a *Messianic realm* in the future of this world [special place associated with anomalous agent].
- One could wish to be saved from being defiled by ritual impurity and hope for the *pure beauty* of psychic and bodily existence [absolute ideal].
- One could wish to escape being incarcerated in an impure body and hope for a *purely spiritual existence* [anomalous experience].
- One could wish to be saved from the eternal and senseless play of human passions and desires and hope for the quietude of the *pure beholding of the divine* [anomalous experience of an anomalous agent].
- One could wish to be saved from radical evil and the servitude of sin and hope for the *eternal and free benevolence* in the lap of *a fatherly god* [ideal action of anomalous agent].
- One could wish to be saved from peonage under the astrologically conceived *determination of stellar constellations* [anomalous non-agent] and long for the dignity of freedom and partaking of the substance of the *hidden deity* [anomalous agent].
- One could wish to be redeemed from the barriers to the finite, which express themselves in suffering, misery and death, and the threatening punishment of *hell*, and hope for *eternal bliss* in an *earthly or paradisiacal future existence* [ideal experience in anomalous place].
- One could wish to be saved from the *cycle of rebirths* [anomalous non-agent] with their inexorable compensations for the deeds of the times past and hope for *eternal rest* [ideal experience].
- One could wish to be saved from senseless brooding and events and long for the *dreamless sleep* [anomalous experience] (Gerth and Mills 1958, 280–81).

Weber's formulation, coupled with his examples, provides a good illustration of a composite formation that stresses the transformative aspect of religions. It was only in the wake of analyzing Weber's formulation that I noticed that I had simplified Buswell and Gimello's (1992) definition of *marga* theory by leaving out the link between practices and transformative power.

Buswell and Gimello. They define *marga* theory as "the theory according to which certain methods of practice, certain prescribed patterns of religious behavior, have transformative power and will lead, somewhat necessarily, to specific religious goals" (Buswell and Gimello 1992, 2–3). Thus, a fuller paraphrase would read: *A religious or spiritual path = a set of* transformative *practices (prescribed patterns of behavior) that individuals or groups view as effective in attaining special goals.* With the addition of the word "transformative," we can more easily compare this formulation of a path deemed religious with Weber's and assess whether the two formulations can be combined and if so how. Schematic variants of transformative practices can probably be worked out using McCauley and Lawson's (2002) work on ritual and Sørensen's (2007) on magic.

Glossary

Ascription—the assignment of a quality or characteristic to some thing.

Ascriptions, simple—ascriptions in which a single thing or event is characterized.

Ascriptions, composite—a set of two or more interlocked ascriptions.

Attributions—the commonsense causal explanations that people offer for why things happen as they do.

Attribution theory—a collection of theories developed by psychologists to explain the commonsense causal explanations that people offer for why things happen as they do.

Consciousness, transitive—consciousness of something.

Consciousness, intransitive—what we are when we are awake, a precondition of consciousness.

Consciousness, first-order—also known as primary or core consciousness, includes sensory awareness, attention, perceptions, memory, emotion, and action. It is present in nonhuman animals and prelinguistic humans and forms the basis for all higher forms of consciousness.

Consciousness, higher-order—includes levels of consciousness beyond first-order consciousness, variously referred to as "conscious awareness," "awareness of awareness," or "meta-awareness." The highest level of consciousness, which is most likely limited to humans, depends upon language and the more complex mental functions associated with it.

Deeming—an umbrella term that encompasses processes of ascription and attribution.

Embodiment—the subjective experience of having and using a body. In cognitive science and the philosophy of mind, it signals an approach that emphasizes the role that the body plays in shaping the mind.

Emic—a term used by anthropologists to refer to language and perspectives of scholars.

Etic—a term used by anthropologists to refer to the language and perspectives of those whom scholars are studying.

Folk psychology—the set of basic, cross-culturally stable assumptions that we use to predict, explain, or understand the everyday actions of others in terms of the mental states we presume lie behind them.

Formations, simple—a simple ascription plus any beliefs and practices associated with it.

Formations, composite—a composite ascription plus the beliefs and practices associated with it.

Magic, magical—treated here as first-order terms and not defined for research purposes.

Mystical—treated here as a first-order term and not defined for research purposes.

Religious—used where convenient as a nontechnical catch-all term for webs of religion-like first-order concepts (for example, "religious," "sacred," "magical," "superstitious," "spiritual," "mystical," as in the phrase "things deemed religious." See "special things" for a more generic, technical, second-order formulation.

Sacred, sacralization—treated here as first-order terms and not defined for research purposes.

Singularization—the processes by means of which people set things apart as special.

Singularities—things considered so special that people set them apart and protect them from other things by means of taboos or prohibitions.

Spiritual—treated here as a first-order term and not defined for research purposes.

Subpersonal—see *unconscious*.

Sui generis—a person or thing that is unique, in a class by itself (Latin, "of its own kind") (Hirsch et al. 2002, s.v. sui generis).

Terminology, first-order—terms that people use in everyday discourse. See *emic*.

Terminology, second-order—terms specifically defined to further scholarly research. See *etic*.

Theory of mind—the folk or everyday understanding of mental states by means of which we predict and explain the behavior of ourselves and others.

Things, special—things that are set apart from other, more ordinary things, including things considered so special that people set them apart as *singularities* and protect them from other things by means of taboos or prohibitions. See also *singularization*.

Things, sets of—temporarily constituted for the purposes of comparison by a stipulated point of analogy.

Unconscious mental activity—refers to any mental activity that takes place below the threshold of consciousness. It is also referred to as the subpersonal level of mental processing and should not be confused with the Freudian unconscious.

Works Cited

Agazzi, Evandro. 1991. *The problem of reductionism in science*. Episteme, 18. Dordrecht: Kluwer Academic Publishers.

Ajilore, O., R. Stickgold, C. D. Rittenhouse, and J. A. Hobson. 1995. Nightcap: Laboratory and home–based evaluation of a portable sleep monitor. *Psychophysiology* 32 (1):92–98.

Allen, Colin, and Mark Bekoff. 2007. Animal consciousness. In *The Blackwell companion to consciousness*, edited by Max Velmans and Susan Schneider. Malden, MA: Blackwell Publishing.

Alles, Gregory D. 2005. Religion [Further Considerations]. *Encyclopedia of Religion*, 2nd ed., edited by Lindsay Jones. Detroit: Macmillan Reference.

American Psychiatric, Association. 1994. *Diagnostic and statistical manual of mental disorders: DSM–IV*. 4th ed. Washington, DC: American Psychiatric Association.

Andresen, Jensine, and Robert K. C. Forman. 2000. Methodological pluralism in the study of religion: How the study of consciousness and mapping spiritual experiences can reshape religious methodology. In *Cognitive models and spiritual maps*, edited by Jensine Andresen and Robert K. C. Forman. Bowling Green, OH: Academic Imprint.

Antaki, Charles. 1988. *Analysing everyday explanation: A casebook of methods*. London: Sage Publications.

Arnal, William E. 2000. Definition. In *Guide to the study of religion*, edited by W. Braun and R. T. McCutcheon. London and New York: Cassell.

Anttonen, Veikko. 1996. Rethinking the sacred: The notions of "human body" and "territory" in conceptualizing religion. In *The sacred and its scholars: Comparative methodologies for the study of primary religious data*, edited by Thomas A. Indinopulos and Edward A. Yonan. Leiden: E. J. Brill.

Anttonen, Veikko. 2000. Space. In *Guide to the study of religion*, edited by W. Braun and R. T. McCutcheon. London and New York: Cassell.

Anttonen, Veikko. 2003. "Sacred sites as markers of difference: Exploring cognitive foundations of territoriality." In *Dynamics of tradition: Perspectives on oral poetry and folk belief*, edited by Lotte Tarkka. Helsinki: Finnish Literature Society.

Arzy, S., G. Thut, C. Mohr, C. M. Michel, and O. Blanke. 2006. Neural basis of embodiment: Distinct contributions of temporoparietal junction and extrastriate body area. *Journal of Neuroscience* 26 (31):8074–81.

Asad, Talal. 1993. The construction of religion as an anthropological category. In *Genealogies of religion: Discipline and reasons of power in Christianity and Islam*, edited by idem. Baltimore: The Johns Hopkins University Press.

Atran, Scott. 2002. *In gods we trust: The evolutionary landscape of religion*. Oxford: Oxford University Press.

Atran, Scott, and Ara Norenzayan. 2004. Religion's evolutionary landscape: Counterintuition, commitment, compassion, communion. *Behavioral and Brain Sciences* 27 (6):713–70.

Azari, Nina P. 2004. Religious experience as thinking that feels like something: A philosophical-theological reflection on recent neuroscientific study of religious experience. *Dissertation Abstracts International Section A: Humanities and Social Sciences*, vol. 65 (2-A):565.

Azari, Nina P., John Missimer, and Rüdiger J. Seitz. 2005. Religious experience and emotion: Evidence for distinctive cognitive neural patterns. *International Journal for the Psychology of Religion* 15 (4):263–81.

Azari, Nina P., Janpeter Nickel, Gilbert Wunderlich, Michael Niedeggen, Harald Hefter, Lutz Tellmann, Hans Herzog, Petra Stoerig, Dieter Birnbacher, and Rüdiger J. Seitz. 2001. Neural correlates of religious experience. *European Journal of Neuroscience* 13 (8):1649–52.

Barkow, Jerome H., Leda Cosmides, and John Tooby. 1992. *The adapted mind: Evolutionary psychology and the generation of culture.* Oxford: Oxford University Press.

Barnard, G. William. 1992. Explaining the unexplainable: Wayne Proudfoot's *Religious Experience. Journal of the American Academy of Religion* 60 (2):231–56.

———. 1997. *Exploring unseen worlds: William James and the philosophy of mysticism.* Albany: State University of New York Press.

———. 1998. William James and the origins of mystical experience. In *The innate capacity: Mysticism, psychology, and philosophy*, edited by R.K.C. Forman. Oxford: Oxford University Press.

Barresi, John, and Chris Moore. 1996. Intentional relations and social understanding. *Behavioral and Brain Sciences* 19 (1):107–54.

Barrett, Justin L. 2004. *Why would anyone believe in God?* Walnut Creek, CA: AltaMira.

———. 2008. Coding and quantifying counterintuitiveness in religious concepts: Theoretical and methodological reflections. *Method and Theory in the Study of Religion* 20 (4):308–38.

Barrett, Justin L., and Frank C. Keil. 1996. Conceptualizing a nonnatural entity: Anthropomorphism in God concepts. *Cognitive Psychology* 31:219–47.

Baumeister, Roy F., ed. 2002. Religion and psychology [Special issue]. *Psychological Inquiry* 13 (3).

Bender, Courtney. 2003. *Heaven's kitchen: Living religion at God's Love We Deliver.* Chicago: University of Chicago Press.

———. 2007. Religion in events. Unpublished paper.

———. 2008. *Worlds of experience: Contemporary spirituality and the American religious imagination.* Chicago: University of Chicago Press.

Bennett, Maxwell, Daniel Dennett, Peter Hacker, and John Searle. 2007. *Neuroscience and philosophy: Brain, mind, and language.* New York: Columbia University Press.

Bennett, Maxwell R., and Peter Hacker. 2003. *Philosophical foundations of neuroscience.* Malden, MA: Blackwell.

———. 2007. The conceptual presuppositions of cognitive neuroscience: A reply to critics. In *Neuroscience and philosophy*, edited by Maxwell Bennett,

Daniel Dennett, Peter Hacker and John Searle. New York: Columbia University Press.

Blanke, Olaf, T. Landis, L. Spinelli, and M. Seeck. 2004. Out-of-body experience and autoscopy of neurological origin. *Brain: A Journal of Neurology* 127 (2):243–58.

Blanke, Olaf, and Christine Mohr. 2005. Out-of-body experience, heautoscopy, and autoscopic hallucination of neurological origin: Implications for neurocognitive mechanisms of corporeal awareness and self-consciousness. *Brain Research Reviews* 50 (1):184–99.

Blanke, Olaf, C. Mohr, C. M. Michel, A. Pascual-Leone, P. Brugger, M. Seeck, T. Landis, and G. Thut. 2005. Linking out-of-body experience and self-processing to mental own-body imagery at the temporoparietal junction. *Journal of Neuroscience* 25 (3):550–57.

Bloom, Paul. 2004. *Descartes' baby: How the science of child development explains what makes us human.* New York: Basic Books.

Boddy, Janice. 1994. Spirit possession revisited: Beyond instrumentality. *Annual Review of Anthropology* 23:407–34.

Bonnell, Victoria E., and Lynn Hunt, eds. 1999. *Beyond the cultural turn: New directions in the study of society and culture.* Berkeley: University of California Press.

Boyer, Pascal. 1994. *The naturalness of religious ideas: A cognitive theory of religion.* Berkeley: University of California Press.

———. 2001. *Religion explained: The evolutionary origins of religious thought.* New York: Basic Books.

Braun, Willi. 2000. Religion. In *Guide to the study of religion,* edited by W. Braun and R. T. McCutcheon. London and New York: Cassell.

Braun, Willi, and Russell T. McCutcheon, eds. 2000. *Guide to the study of religion.* London and New York: Cassell.

Brewer, Marilynn B. 2003. *Intergroup relations.* 2nd ed. Philadelphia, PA: Open University Press.

Brown, Rupert. 2000a. Social identity theory: Past achievements, current problems and future challenges. *European Journal of Social Psychology* 30 (6):745–78.

———. 2000b. *Group processes: Dynamics within and between groups.* 2nd ed. Oxford: Blackwell.

Brown, Terrance, and Leslie Smith, eds. 2003. *Reduction and the development of knowledge.* Mahwah, NJ: Erlbaum.

Buhrman, Sarasvati. 1997. Trance types and amnesia revisited: Using detailed interviews to fill in the gaps. *Anthropology of Consciousness* 8:10–21.

Buswell, Robert E., and Robert M. Gimello, eds. 1992. *Paths to liberation: The mārga and its transformations in Buddhist thought.* Honolulu: University of Hawaii Press.

Bynum, Caroline Walker. 1987. *Holy feast and holy fast: The religious significance of food to medieval women.* Berkeley: University of California Press.

———. 1992. *Fragmentation and redemption: Essays on gender and the human body in medieval religion.* New York: Zone Books.

Caciola, Nancy. 2003. *Discerning spirits: Divine and demonic possession in the Middle Ages.* Ithaca: Cornell University Press.

Cacioppo, John T., and Gary G. Berntson. 2005. *Social neuroscience: Key readings*. New York: Psychology Press.

Cahn, B. R., and J. Polich. 2006. Meditation states and traits: EEG, ERP, and neuroimaging studies. *Psychological Bulletin* 132 (2):180–211.

Camille, Michael. 1996. *Gothic art: Glorious visions*. Upper Saddle River, NJ: Prentice-Hall.

Cannell, Fenella. 2006. *The anthropology of Christianity*. Durham: Duke University Press.

Capps, Walter H. 1995. *Religious studies: The making of a discipline*. Minneapolis: Fortress.

Cardeña, Etzel, Steven Jay Lynn, and Stanley Krippner. 2000. *Varieties of anomalous experience: Examining the scientific evidence*. Washington, DC: American Psychological Association.

Carter, Jeffrey. 2004. Comparison in the history of religions: Reflections and critiques. *Method and Theory in the Study of Religion* 16 (1):3–11.

Certeau, Michel de. 1995. *The mystic fable: Religion and postmodernism*. Chicago: University of Chicago Press.

Cheyne, J. Allen. 2000. Play, dreams, and simulation (A brief commentary on Revonsuo). *Behavioral and Brain Sciences* 23:918–19.

———. 2001. The ominous numinous: Sensed presence and "other" hallucinations. In *Between ourselves: Second-person issues in the study of consciousness*, edited by E. Thompson. Charlottesville, VA: Imprint Academic.

———. 2003. Sleep paralysis and the structure of waking-nightmare hallucinations. *Dreaming* 13 (3):163–79.

Cheyne, J. Allen, and Todd A. Girard. 2007a. Paranoid delusions and threatening hallucinations: A prospective study of sleep paralysis experiences. *Consciousness and Cognition* 16 (4):959–74.

———. 2007b. The nature and varieties of felt presence experiences: A reply to Nielsen. *Consciousness and Cognition* 16 (4):984–91.

Cheyne, J. Allen, Steve D. Rueffer, and Ian R. Newby-Clark. 1999. Hypnagogic and hypnopompic hallucinations during sleep paralysis: Neurological and cultural construction of the nightmare. *Consciousness and Cognition* 8 (3):319–37.

Chidester, David. 1996. *Savage systems: Colonialism and comparative religion in southern Africa*. Charlottesville, VA: University of Virginia Press.

Chidester, David, and Edward T. Linenthal. 1995. *American sacred space*. Bloomington and Indianapolis: Indiana University Press.

Chiu, Chi-Yue, and Ying-Yi Hong. 2006. *Social psychology of culture*. New York: Psychology Press.

Christian, William A. 1992. *Moving crucifixes in modern Spain*. Princeton: Princeton University Press.

Cicogna, P., and M. Bosinelli. 2001. Consciousness during dreams. *Consciousness and Cognition* 10 (1):26–41.

Clancy, Susan A. 2005. *Abducted: How people come to believe they were kidnapped by aliens*. Cambridge, MA: Harvard University Press.

Clayton, Philip. 2004. *Mind and emergence: From quantum to consciousness*. Oxford: Oxford University Press.

Clayton, Philip, and Paul Davies. 2006. *The re-emergence of emergence: The emergentist hypothesis from science to religion*. Oxford: Oxford University Press.

Cohen, Emma. 2008. What is spirit possession? Defining, comparing, and explaining two possession forms. *Ethnos* 73 (1):101–26.

Cox, James L. 2006. *A guide to the phenomenology of religion: Key figures, formative influences and subsequent debates*. London: Clark.

Csikszentmihalyi, Mihaly, and Eugene Rochberg-Halton. 1981. *The meaning of things: Domestic symbols and the self*. New York: Cambridge University Press.

Csordas, Thomas. 1994. *Embodiment and experience: The existential ground of culture and self*. New York: Cambridge University Press.

D'Aquili, Eugene G., and Andrew B. Newberg. 1999. *The mystical mind: Probing the biology of religious experience*. Minneapolis: Fortress.

Damasio, Antonio R. 1994. *Descartes' error: Emotion, reason, and the human brain*. New York: G. P. Putnam.

———. 1999. *The feeling of what happens: Body and emotion in the making of consciousness*. New York: Harcourt Brace.

Daston, Lorraine. 1991. Marvelous facts and miraculous evidence in early modern Europe. *Critical Inquiry* 18 (1):93–124.

Davis, Caroline Franks. 1989. *The evidential force of religious experience*. Oxford: Oxford University Press.

Day, Matthew. 2005. The undiscovered and undiscoverable essence: Species and religion after Darwin. *Journal of Religion* 85 (1):58–82.

De Rivera, Joseph, and Theodore R. Sarbin. 1998. *Believed-in imaginings: The narrative construction of reality*. Washington, DC: American Psychological Association.

De Vries, Hent, ed. 2008. *Religion: Beyond a concept*. New York: Fordham University Press.

Deacon, Terrence William. 1997. *The symbolic species: The co-evolution of language and the brain*. New York: W.W. Norton.

Dennett, Daniel C. 1976. Are dreams experiences? *Philosophical Review* 85 (2):151–71.

———. 1991. *Consciousness explained*. Boston: Little, Brown.

———. 1993. Review of Varela et al., The embodied mind. Available online at http://www.ase.tufts.edu/cogstud/papers/varela.htlm.

———. 2003. Who's on first? Heterophenomenology explained. *Journal of Consciousness Studies* 10 (9–10):19–30.

———. 2007a. Philosophy as naive anthropology: Comment on Bennett and Hacker. In *Neuroscience and philosophy*, edited by Maxwell Bennett, Daniel Dennett, Peter Hacker and John Searle. New York: Columbia University Press.

———. 2007b. Heterophenomenology reconsidered. *Phenomenology and Cognitive Science* 6:247–70.

Dewey, John. 1934. *A common faith*. New Haven: Yale University Press.

Dick, Michael Brennan, ed. 1999. *Born in heaven, made on earth: The creation of the cult image in the ancient near east*. Winona Lake, IN: Eisenbrauns.

Dittes, James L. 1969. Psychology of religion. In *The handbook of social psychology*, vol. 5, edited by G. Lindzey and E. Aronson. Reading, MA: Addison-Wesley.

Dow, James W. 2007. A scientific definition of religion. *Anpere: Anthropological Perspectives on Religion.* Available online at http://.www.anpere.net.

Durkheim, Emile. 1912/1995. *The elementary forms of religious life.* Translated by Karen Fields. New York: Free Press.

———. 1912/2001. *The elementary forms of religious life.* Translated by Carol Cosman. Edited by Mark S. Cladis. Oxford: Oxford University Press.

Edelman, Gerald M. 1992. *Bright air, brilliant fire: On the matter of the mind.* New York: Basic Books.

Ehrsson, H. Henrik. 2007. The experimental induction of out-of-body experiences. *Science* 317 (5841):1048.

Eliade, Mircea. 1957/1987. *The sacred and the profane: The nature of religion.* Translated by W. R. Trask. New York: Harcourt.

Emmons, Robert A., and Raymond F. Paloutzian. 2003. The psychology of religion. *Annual Review of Psychology* 54:377–402.

Fauconnier, Gilles, and Mark Turner. 2002. *The way we think: Conceptual blending and the mind's hidden complexities.* New York: Basic Books.

Fiske, Alan Page, and Philip E. Tetlock. 1997. Taboo trade-offs: Reactions to transactions that transgress the spheres of justice. *Political Psychology* 18 (2):255–97.

Fiske, Susan T., and Shelley E. Taylor. 1991. *Social cognition,* 2nd ed. New York: McGraw-Hill.

Fitzgerald, Timothy. 2000a. *The ideology of religious studies.* New York: Oxford University Press.

———. 2000b. Experience. In *Guide to the study of religion,* edited by W. Braun and R. T. McCutcheon. London and New York: Cassell.

———. 2007a. *Discourse on civility and barbarity: A critical history of religion and related categories.* New York: Oxford University Press.

———, ed. 2007b. *Religion and the secular: Historical and colonial formations.* London: Equinox.

Flanagan, Owen J. 2000. *Dreaming souls: Sleep, dreams, and the evolution of the conscious mind.* Oxford: Oxford University Press.

———. 2007. *The really hard problem: Meaning in a material world.* Cambridge, MA: The MIT Press.

Forman, Robert K. C. 1990. *The problem of pure consciousness: Mysticism and philosophy.* New York: Oxford University Press.

———. 1998. *The innate capacity: Mysticism, psychology, and philosophy.* New York: Oxford University Press.

———. 1999. *Mysticism, mind, consciousness.* Albany: State University of New York Press.

———. 2008. Neuroscience, Consciousness and Spirituality Conference, July 2–4, 2008, Freiburg Germany. *Journal of Consciousness Studies* 15 (8):110–15.

Fosse, R., R. Stickgold, and J. Allan Hobson. 2001. Brain-mind states: Reciprocal variation in thoughts and hallucinations. *Psychological Science* 12 (1):30–36.

———. 2004. Thinking and hallucinating: Reciprocal changes in sleep. *Psychophysiology* 41 (2):298–305.

Foster, Genevieve W., and David Hufford. 1985. *The world was flooded with light: A mystical experience remembered.* Pittsburgh, PA: University of Pittsburgh Press.

Foulkes, David. 1985. *Dreaming: A cognitive-psychological analysis.* Hillsdale, NJ: Erlbaum.

Frigerio, A. 1989. Levels of possession awareness in Afro-Brazilian religions. *Association for the Anthropological Study of Consciousness* 5 (2–3):5–11.

Försterling, Friedrich. 2001. *Attribution: An introduction to theories, research, and applications.* Philadelphia: Psychology Press.

Gallagher, Shaun. 2001. The practice of mind: Theory, simulation or interaction? In *Between ourselves: Second-person issues in the study of consciousness*, edited by Evan Thompson. Charlottesville, VA: Imprint Academic.

———. 2003. Phenomenology and experimental design: Toward a phenomenologically enlightened experimental science. *Journal of Consciousness Studies* 10 (9–10):85–99.

———. 2005. *How the body shapes the mind.* New York: Oxford University Press.

———. 2007. Phenomenological approaches to consciousness. In *The Blackwell companion to consciousness*, edited by M. Velmans and S. Schneider. Malden, MA: Blackwell.

Gallagher, Shaun, and Daniel Hutto. 2008. Understanding others through primary interaction and narrative practice. In *The shared mind: Perspectives on intersubjectivity*, edited by J. Zlatev, T. Racine, C. Sinha and E. Itkonen. Amsterdam: John Benjamins.

Gallagher, Shaun, and Jesper Sørensen. 2006. Experimenting with phenomenology. *Consciousness and Cognition* 15 (1):119–34.

Gallese, Vittorio, and George Lakoff. 2005. The brain's concepts: The role of the sensory-motor system in conceptual knowledge. *Cognitive Neuropsychology* 22 (3–4):455–79.

Geary, Patrick. 1986. Sacred commodities: The circulation of medieval relics. In *The social life of things: Commodities in cultural perspective*, edited by Arjun Appadurai. Cambridge: Cambridge University Press.

Geertz, Clifford. 1973. *The interpretation of cultures.* New York: Basic Books.

Gerth, H. H., and C. Wright Mills, eds. 1958. *From Max Weber: Essays in sociology.* New York: Oxford University Press.

Gibbs, Raymond W. 2006. *Embodiment and cognitive science.* New York: Cambridge University Press.

Glaser, Jack, and John F. Kihlstrom. 2005. Compensatory automaticity: Unconscious volition is not an oxymoron. In *The new unconscious*, edited by Ran R. Hassin, James S. Uleman, and John A. Bargh. New York: Oxford University Press.

Glass, Matthew. 1995. "Alexanders All": Symbols of conquest and resistance at Mount Rushmore. In *American sacred space*, edited by David Chidester and Edward T. Linenthal. Bloomington and Indianapolis: Indiana University Press.

Graeber, David. 2001. *Toward an anthropological theory of value: The false coin of our own dreams.* New York: Palgrave.

Haidt, Jonathan. 2000. The positive emotion of elevation. *Prevention and Treatment.* Available online at http://journals.apa.org/prevention/volume3/pre 003003c.html.

———. 2003. The moral emotions. In *Handbook of affective sciences*, edited by R. J. Davidson, K. R. Scherer, and H. H. Goldsmith. Oxford: Oxford University Press.

———. 2007. The new synthesis in moral psychology." *Science* 316:998–1002.

Haidt, Jonathan, and Craig Joseph. 2004. Intuitive ethics: How innately prepared intuitions generate culturally variable virtues." *Daedalus* 133 (4): 55–66.

Halbfass, Wilhelm. 1988. The concept of experience in the encounter between India and the West. In *India and Europe: An essay in understanding*, edited by idem. Albany: State University of New York Press.

Halperin, Daniel. 1995. Memory and "consciousness" in an evolving Brazilian possession religion. *Anthropology of Consciousness* 6:1–17.

———. 1996a. A delicate science: A critique of an exclusively emic anthropology. *Anthropology and Humanism* 21:31–40.

———. 1996b. Trance and possession: Are they the same? *Transcultural Psychiatric Research Review* 33 (1):33–41.

Hanegraaff, Wouter J. 2003. How magic survived the disenchantment of the world. *Religion* 33 (4):357–80.

Harris, Ruth. 1999. *Lourdes: Body and spirit in the secular age*. New York: Penguin Compass.

Hassin, Ran R., James S. Uleman, and John A. Bargh. 2005. *The new unconscious*. Oxford: Oxford University Press.

Heap, Michael, Richard J. Brown, and David A. Oakley. 2004. *The highly hypnotizable person: Theoretical, experimental and clinical studies*. London: Routledge.

Hervieu-Léger, Danièle. 2000. *Religion as a chain of memory*. New Brunswick, NJ: Rutgers University Press.

Hewstone, Miles. 1989. *Causal attribution: From cognitive processes to collective beliefs*. Cambridge, MA: Blackwell.

———. 1995. Attribution theories. In *The Blackwell encyclopedia of social psychology*, edited by A.S.R. Manstead and M. Hewstone. Oxford: Blackwell.

Hirsch, E.D. Jr., Joseph F. Kett, and James Trefil, eds. 2002. *The new dictionary of cultural literacy*, 3rd ed. New York: Houghton Mifflin.

Hishikawa, Y., and T. Shimizu. 1995. Physiology of REM sleep, cataplexy, and sleep paralysis. *Advances in Neurology* 67:245–71.

Hobson, J. Allen. 1999. *Consciousness*. New York: Scientific American Library.

———. 2001. *The dream drugstore: Chemically altered states of consciousness*. Cambridge, MA: The MIT Press.

———. 2007a. Normal and abnormal states of consciousness. In *The Blackwell companion to consciousness*, edited by Max Velmans and Susan Schneider. Malden, MA: Blackwell.

———. 2007b. States of consciousness: Normal and abnormal variation. In *The Cambridge handbook of consciousness*, edited by Philip David Zelazo, Morris Moscovitch, and Evan Thompson. Cambridge: Cambridge University Press.

Hobson, J. Allen, Edward. F. Pace-Schott, and R. Stickgold. 2000a. Dreaming and the brain: Toward a cognitive neuroscience of conscious states. *Behavioral and Brain Sciences* 23 (6):793–842.

———. 2000b. Dream science 2000: A response to commentaries on "Dreaming and the brain." *Behavioral and Brain Sciences*. Special Issue: *Sleep and dreaming* 23 (6):1019–35; 1083–1121.

Hobson, J. Allen, E. F. Pace-Schott, R. Stickgold, and D. Kahn. 1998. To dream or not to dream? Relevant data from new neuroimaging and electrophysiological studies. *Current opinion in neurobiology* 8 (2):239–44.

Hogg, Michael A., and Dominic Abrams. 2001. *Intergroup relations: Essential readings, Key readings in social psychology*. Philadelphia: Psychology Press.

Holdrege, Barbara A. 1995. *Veda and Torah: Transcending the textuality of scripture*. Albany: State University of New York Press.

Hood, Ralph W. 2005. Mystical, spiritual, and religious experiences. In *Handbook of the Psychology of Religion and Spirituality*, edited by R. F. Paloutzian and C. L. Park. New York: Guilford.

———. 2006. The common core thesis in the study of mysticism. In *Where God and science meet,* vol. 3: *The psychology of religious experience*, edited by Patrick McNamara. Westport, CT: Praeger.

Hufford, David J. 1982. *The terror that comes in the night: An experience-centered study of supernatural assault traditions*. Philadelphia: University of Pennsylvania Press.

———. 2005. Sleep paralysis as spiritual experience. *Transcultural Psychiatry* 42 (1):11–45.

Hurford, James R. 2007. *The origins of meaning: Language in light of evolution*. New York: Oxford University Press.

Hutchins, Edwin. 1995. *Cognition in the wild*. Cambridge, MA: The MIT Press.

Hutto, Daniel D. 2007a. First communions: Mimetic sharing without theory of mind. In *The shared mind: Perspectives on intersubjectivity*, edited by J. Zlatev, T. Racine, C. Sinha, and E. Itkoken. Amsterdam: John Benjamins.

———. 2007b. Folk psychology without theory or simulation. In *Folk psychology reassessed*, edited by D. D. Hutto and M. Ratcliffe. Dordrecht: Springer.

———. 2007c. The narrative practice of hypnosis. In *Narrative and understanding persons*, edited by D. D. Hutto. Cambridge, MA: Cambridge University Press.

———. 2008. *Folk psychological narratives: The sociocultural basis of understanding reasons*. Cambridge, MA: The MIT Press.

Huxley, Aldous. 1944. *The perennial philosophy*. New York: Harper Colophon.

James, William. 1902/1985. *The varieties of religious experience*. Cambridge, MA: Harvard University Press.

———. 1904/1987. A world of pure experience. In *William James: Writings 1902–1910*. New York: Library of America.

———. 1992–2004. *The Correspondence of William James*, 12 vols. Edited by Ignas K. Skrupskelis and Elizabeth M. Berkeley. Charlottesville: University of Virginia Press.

Jantzen, Grace M. 1990. Could there be a mystical core of religion? *Religious Studies* 26 (1):59–71.

———. 1995. *Power, gender, and Christian mysticism*. Cambridge: Cambridge University Press.

Jay, Martin. 2005. *Songs of experience: Modern American and European variations on a universal theme*. Berkley: University of California Press.

Jensen, Jeppe Sinding. 2003. *The study of religion in a new key: Theoretical and philosophical soundings in the comparative and general study of religion.* Aarhus: Aarhus University Press.

Johnson, Mark. 1987. *The body in the mind: The bodily basis of meaning, imagination, and reason.* Chicago: University of Chicago Press.

———. 1991. "Knowing through the body." *Philosophical Psychology* 4:3–18.

Kabat-Zinn, Jon. 1982. An outpatient program in behavioral medicine for chronic pain patients based on the practice of mindfulness meditation: Theoretical considerations and preliminary results. *General Hospital Psychiatry* 4 (1):33–47.

———. 2005. *Full catastrophe living: Using the wisdom of your body and mind to face stress, pain, and illness.* 15th anniversary ed. New York: Bantam Dell.

Kahan, Tracy L., and Stephen LaBerge. 1994. Lucid dreaming as metacognition: Implications for cognitive science. *Consciousness and Cognition* 3 (2):246–64.

Kallio, Sakari, and Antti Revonsuo. 2003. Hypnotic phenomena and altered states of consciousness: A multilevel framework of description and explanation. *Contemporary Hypnosis* 20 (3):111–64.

Kandinsky, Wassily. 1911/2006. *Concerning the spiritual in art.* Boston: MFA Publications.

Katz, Steven T. 1978. *Mysticism and philosophical analysis.* New York: Oxford University Press.

———. 1983. *Mysticism and religious traditions.* New York: Oxford University Press.

Kaufman, Suzanne. 2005. *Consuming visions: Mass culture and the Lourdes shrine.* Ithaca: Cornell University Press.

Kelly, Edward F., Emily Williams Kelly, Adam Crabtree, Alan Gauld, Michael Grosso, and Bruce Greyson. 2007. *Irreducible mind: Towards a psychology for the twenty-first century.* Lanham, MD: Rowan and Littlefield.

Keltner, Dacher, and Jonathan Haidt. 2003. Approaching awe, a moral, spiritual, and aesthetic emotion. *Cognition and Emotion* 17 (2):297–314.

Kenny, Michael G. 1981. Multiple personality and spirit possession. *Psychiatry: Journal for the Study of Interpersonal Processes* 44 (4):337–58.

Kim, Jaegwon. 2006. Being realistic about emergence. In *The re-emergence of emergence: The emergentist hypothesis from science to religion,* edited by P. Clayton and P. Davies. New York: Oxford University Press.

King, Richard. 1999. *Orientalism and religion: Postcolonial theory, India and "the mystic East."* London: Routledge.

King, Winston. 1987. Religion. In *Encyclopedia of Religion,* edited by Mircea Eliade. New York: Macmillan.

Kirkpatrick, Lee A. 2005a. *Attachment, evolution, and the psychology of religion.* New York: Guilford.

———. 2005b. Evolutionary psychology: An emerging new foundation for the psychology of religion. In *Handbook of the psychology of religion and spirituality,* edited by R. F. Paloutzian and C. L. Park. New York: Guilford Press.

Klaniczay, Gabor. 2007. The process of trance: Heavenly and diabolic apparitions in Johannes Nider's *Formicarius.* In *Procession, performance, liturgy, and ritual*, edited by Nancy van Deusen. Ottawa: Claremont Cultural Studies, Wissenschaftliche Abhandlungen Bd. 62 (8).

Knott, Kim. 2005. *The location of religion: A spatial analysis*. London: Equinox.

Kopytoff, Igor. 1986. The cultural biography of things: Commoditization as process. In *The social life of things: Commodities in cultural perspective*, edited by Arjun Appadurai. Cambridge: Cambridge University Press.

Kripal, Jeffrey John. 2001. *Roads of excess, palaces of wisdom: Eroticism and reflexivity in the study of mysticism*. Chicago: University of Chicago Press.

———. 2007. *The serpent's gift: Gnostic reflections on the study of religion*. Chicago: University of Chicago Press.

Kroll, Jerome, and Bernard Bachrach. 2005. *The mystic mind: The psychology of medieval mystics and ascetics*. New York: Routledge.

Kselman, Thomas A. 1983. *Miracles and prophecies in nineteenth-century France*. New Brunswick, NJ: Rutgers University Press.

LaBerge, Stephen. 1985. *Lucid dreaming*. Los Angeles: J.P. Tarcher.

———. 2000. Lucid dreaming: Evidence and methodology. *Behavioral and Brain Sciences*. Special Issue: *Sleep and dreaming* 23:962–64.

Lakoff, George, and Mark Johnson. 1980. *Metaphors we live by*. Chicago: University of Chicago Press.

———. 1989. *Philosophy in the flesh*. New York: Cambridge University Press.

Lawson, E. Thomas, and Robert N. McCauley. 1990. *Rethinking religion: Connecting cognition and culture*. Cambridge: Cambridge University Press.

Leeuw, G. van der. 1933/1986. *Religion in essence and manifestation*. Princeton: Princeton University Press.

Lenggenhager, Bigna, Tej Tadi, Thomas Metzinger, and Olaf Blanke. 2007. *Video ergo sum*: Manipulating bodily self-consciousness. *Science* 317 (5841): 1096–99.

Liberman, Debra, John Tooby, and Leda Cosmides. 2007. The architecture of human kin detection. *Nature* 445.

Linenthal, Edward T. 1995. Locating holocaust memory: The United States Holocaust Memorial Museum. In *American sacred space*, edited by David Chidester and Edward T. Linenthal. Bloomington and Indianapolis: Indiana University Press.

Livingston, Kenneth R. 2005. Religious practice, brain, and belief. *Journal of Cognition and Culture* 5 (1–2):75–117.

Loftus, Elizabeth F. 1979. *Eyewitness testimony*. Cambridge, MA: Harvard University Press.

Luhrmann, Tanya M. 1989. *Persuasions of the witch's craft: Ritual magic in contemporary England*. Cambridge, MA: Harvard University Press.

———. 2004. Yearning for God: Trance as a culturally specific practice and its implications for understanding dissociative disorders. *Journal of Trauma and Dissociation*. Special Issue: *Dissociation in Culture* 5 (2):101–29.

———. 2005. The art of hearing God: Absorption, dissociation, and contemporary American spirituality. *Spiritus* 5 (2):133–57.

———. Forthcoming. Feeling the force: Making sense of raw moments in fieldwork. In *Emotion in the Field*, edited by James Davies. Stanford: Stanford University Press.

Luther, Martin. 1520. On the Babylonian captivity of the church. In *Project Wittenberg Online Electronic Study Edition*, edited by Robert E. Smith. Available online at http://www.ctsfw.edu/etext/luther/babylonian/babylonian.htm.

Lutz, Antoine, John D. Dunne, and Richard J. Davidson. 2007. Meditation and the neuroscience of consciousness: An introduction. In *The Cambridge handbook of consciousness,* edited by Philip David Zelazo, Morris Moscovitch, and Evan Thompson. Cambridge: Cambridge University Press.

Mageo, Jeannette Marie, and Alan Howard. 1996. *Spirits in culture, history, and mind.* New York: Routledge.

Mahony, A., K. Pargament, A. Murray-Schwank, and N. Murray-Schwank. 2003. Religion and the sanctification of family relationships. *Review of Religious Research* 44:220–36.

Malcolm, N. 1956. Dreaming and skepticism. *Philosophical Review* 65 (1):14–37.

Malle, Bertram F. 2004. *How the mind explains behavior: Folk explanations, meaning, and social interaction.* Cambridge, MA: The MIT Press.

———. 2005a. Folk theory of mind: Conceptual foundations of human social cognition. In *The new unconscious,* edited by R. Hasin, J. S. Uleman, and J. A. Bargh. New York: Oxford University Press.

———. 2005b. The world and words of mind. *Psychological Inquiry* 16 (1): 21–26.

———. 2006. Of windmills and straw men: Folk assumptions of mind and action. In *Does consciousness cause behavior?,* edited by S. Pockett, W. P. Banks, and S. Gallagher. Cambridge, MA: The MIT Press.

Mamelak, A., and J. A. Hobson. 1989. Nightcap: A home-based sleep-monitoring system. *Sleep* 12 (2):157–66.

Marsden, George M., and Bradley J. Longfield. 1992. *The secularization of the academy.* New York: Oxford University Press.

Martin, Luther H. 2004. "Disenchanting" the comparative study of religion. *Method and Theory in the Study of Religion* 16 (1):36–44.

Masuzawa, Tomiko. 2005. *The invention of world religions.* Chicago: University of Chicago Press.

Mayaram, Shail. 2001. Recent anthropological work on spirit possession. *Religious Studies Review* 27 (3):213–22.

McCauley, Robert N., and E. Thomas Lawson. 2002. *Bringing ritual to mind: Psychological foundations of cultural forms.* Cambridge: Cambridge University Press.

McClenon, James. 1994. *Wondrous events: Foundations of religious belief.* Philadelphia: University of Pennsylvania Press.

McCutcheon, Russell T. 1997. *Manufacturing religion: The discourse on sui generis religion and the politics of nostalgia.* New York: Oxford University Press.

McDannell, Colleen. 1995. *Material Christianity: Religion and popular culture in America.* New Haven: Yale University Press.

McGinn, Bernard. 1998. *The flowering of mysticism: Men and women in the new mysticism (1200–1350).* New York: Crossroad.

McNally, R. J., and S. A. Clancy. 2005a. Sleep paralysis, sexual abuse, and space alien abduction. *Transcultural Psychiatry* 42 (1):113–22.

———. 2005b. Sleep paralysis in adults reporting repressed, recovered, or continuous memories of childhood sexual abuse. *Journal of Anxiety Disorders* 19:595–602.

Merikle, Phil. 2007. Preconscious processing. In *The Blackwell companion to consciousness*, edited by Max Velmans and Susan Schneider. Malden, MA: Blackwell.

Merkur, Daniel. 2000. Review of Robert K. C. Forman, *Mysticism, mind, consciousness. Religion* 30:405–7.

Monroe, John Warne. 2008. *Laboratories of faith: Mesmerism, spiritism, and occultism.* Ithaca: Cornell University Press.

Murphy, Nancy. 2006. Emergence and mental causation. In *The re-emergence of emergence: The emergentist hypothesis from science to religion*, edited by P. Clayton and P. Davies. New York: Oxford University Press.

Nagel, Thomas. 1986. *The view from nowhere.* New York: Oxford University Press.

Nanda, Meera. 2003. *Prophets facing backward: Postmodern critiques of science and Hindu nationalism in India.* New Brunswick, NJ: Rutgers University Press.

Nelson, K. R., M. Mattingly, S. A. Lee, and F. A. Schmitt. 2006. Does the arousal system contribute to near-death experience? *Neurology* 66 (7):1003–9.

Nelson, K. R., M. Mattingly, and F. A. Schmitt. 2007. Out-of-body experience and arousal. *Neurology* 68 (10):794–95.

Newberg, Andrew, Abass Alavi, Michael Baime, Michale Pourdehnad, Jill Santanna, Eugene d'Aquili. 2001a. The measurement of regional cerebral blood flow during the complex cognitive task of meditation: A preliminary SPECT study. *Psychiatry Research: Neuroimaging Section* 106:113–22.

Newberg, Andrew B., Eugene G. d'Aquili, Vince Rause, and Judith Cummins. 2001b. *Why God won't go away: Brain science and the biology of belief.* New York: Ballantine.

Newman, Barbara. 2005. What did it mean to say "I saw"? The clash between theory and practice in medieval visionary culture. *Speculum* 80 (1):1–43.

Nielsen, T. 2007. Felt presence: Paranoid delusion or hallucinatory social imagery? *Consciousness and Cognition* 16 (4):975–83.

Noll, Richard, Jr. 1985. Mental imagery cultivation as a cultural phenomenon: The role of visions in shamanism. *Current Anthropology* 26 (4):443–61.

Oohashi, T, N. Kawai, M. Honda, S. Nakamura, M. Morimoto, E. Nishina, and T. Maekawa. 2002. Electroencephalographic measurement of possession trance in the field. *Clinical Neurophysiology* 113 (3):435–45.

Otto, Rudolf. 1923/1958. *The idea of the holy.* London: Oxford University Press.

Pace-Schott, Edward F. 2003. *Sleep and dreaming: Scientific advances and reconsiderations.* Cambridge: Cambridge University Press.

Pace-Schott, Edward F., and J. Allan Hobson. 2007. Altered states of consciousness: drug-induced states. In *The Blackwell companion to consciousness*, edited by Max Velmans and Susan Schneider. Malden, MA: Blackwell.

Paciotti, Brian, Peter J. Richerson, and Robert Boyd. 2006. Cultural evolutionary theory: A synthetic theory for fragmented disciplines. In *Bridging Social*

Psychology: The Benefits of Transdisciplinary Approaches, edited by P. Van Lange. Mahwah, N.J.: Erlbaum.

Paden, William. 1994. Before "the sacred" became theological: Durkheim and reductionism. In *Religion and reductionism: Essays on Eliade, Segal, and the challenge of the social sciences for the study of religion*, edited by T. A. Idinopulos and E. A. Yonan. Leiden: Brill.

———. 2001. Universals revisited: Human behaviors and cultural variations. *Numen* 48 (3):276–89.

———. 2005. Comparative religion. In *Encyclopedia of religion*, 2nd ed. New York: Macmillan.

Paloutzian, Raymond F. 2005. Religious conversion and spiritual transformation: A meaning-systems analysis. In *Handbook of the psychology of religion and spirituality*, edited by R. F. Paloutzian and C. L. Park. New York: Guilford.

Paloutzian, Raymond F., and Crystal L. Park. 2005a. Integrative themes in the current science of the psychology of religion. In *Handbook of the psychology of religion and spirituality*, edited by R. F. Paloutzian and C. L. Park. New York: Guilford.

———, eds. 2005b. *Handbook of the psychology of religion and spirituality*. New York: Guilford.

Pals, Daniel L. 1986. Reductionism and belief: An appraisal of recent attacks on the doctrine of irreducible religion. *Journal of Religion* 66 (1):18–36.

———. 1987. Is religion a sui generis phenomenon? *Journal of the American Academy of Religion* 55 (2):259–82.

Panksepp, Jaak. 1998. *Affective neuroscience: The foundations of human and animal emotions*. New York: Oxford University Press.

———. 2007. Affective consciousness. In *The Blackwell companion to consciousness*, edited by Max Velmans and Susan Schneider. Malden, MA: Blackwell.

Panzarasa, Pietro, and Nicholas R. Jennings. 2006. Collective cognition and emergence in multi-agent systems. In *Cognition and multi-agent interaction: From cognitive modeling to social simulation*, edited by R. Sun. Cambridge: Cambridge University Press.

Pargament, Kenneth I. 1997. *The psychology of religion and coping: Theory, research, practice*. New York: Guilford.

———. 2002. Is religion nothing but . . . ? Explaining religion versus explaining religion away. *Psychological Inquiry* 13 (3):239–44.

Pargament, Kenneth I., Gina M. Magyar-Russell, and Nichole A. Murray-Schwank. 2005. The sacred and the search for significance: Religion as a unique process. *Journal of Social Issues* 61 (4):665–87.

Pargament, Kenneth I., and Annette Mahoney. 2005. Sacred matters: Sanctification as a vital topic for the psychology of religion. *International Journal for the Psychology of Religion*. 15 (3):179–98.

Park, Crystal L. 2005a. Religion and meaning. In *Handbook of the psychology of religion and spirituality*, edited by R. F. Paloutzian and C. L. Park. New York: Guilford.

———. 2005b. Religion as a meaning-making framework in coping with life stress. *Journal of Social Issues* 61 (4):707–29.

Park, Crystal L., and Susan Folkman. 1997. Meaning in the context of stress and coping. *Review of General Psychology* 1 (2):115–44.

Pelikan, Jaroslav. 1983. *Reformation of church and dogma (1300–1700)*. Chicago: University of Chicago Press.

Phillips, R. E., and Kenneth I. Pargament. 2002. The sanctification of dreams: Prevalence and implications. *Dreaming* 12 (3):141–53.

Pockett, Susan, William P. Banks, and Shaun Gallagher, eds. 2006. *Does consciousness cause behavior?* Cambridge, MA: The MIT Press.

Poole, Fitz Porter. 1986. Metaphors and maps: Towards comparison in the anthropology of religion. *Journal of the American Academy of Religion* 54 (3): 411–57.

Proudfoot, Wayne. 1985. *Religious experience*. Berkeley: University of California Press.

———. 1993. Explaining the unexplainable. *Journal of the American Academy of Religion* 61 (4):793–803.

Proudfoot, Wayne, and Phillip Shaver. 1975. Attribution theory and the psychology of religion. *Journal for the Scientific Study of Religion* 14 (4):317–30.

Pullella, Philip. 2008 (May 16). Pope restates gay marriage ban after California vote. Thompson, Reuters. Available online at http://www.reuters.com/article/idUSL1627550020080516.

Pyysiäinen, Ilkka. 2003a. *How religion works: Towards a new cognitive science of religion. Cognition and culture series*. Leiden: Brill.

———. 2003b. Counterintuitiveness as the hallmark of religiosity. *Religion* 33: 341–55.

———. 2004. *Magic, miracles, and religion: A scientist's perspective. Cognitive science of religion series*. Walnut Creek, CA: AltaMira.

Rappaport, Roy A. 1999. *Ritual and religion in the making of humanity*. Cambridge: Cambridge University Press.

Rennie, Bryan S. 1996. *Reconstructing Eliade: Making sense of religion*. Albany: State University of New York Press.

Revonsuo, Antti. 2000. The reinterpretation of dreams: An evolutionary hypothesis of the function of dreaming. *Behavioral and Brain Sciences*. Special Issue: *Sleep and dreaming* 23:793–1121.

Richerson, Peter, and Robert Boyd. 2001. Culture is part of human biology: Why the superorganic concept serves the human sciences badly. In *Science studies: Probing the dynamics of scientific knowledge*, edited by S. Maasen and M. Winterhager. Bielefeld: Verlag.

Robinson, Daniel. 2007. Still looking: Science and philosophy in pursuit of prince reason. In *Neuroscience and philosophy*, edited by M. Bennett, D. Dennett, P. Hacker, and J. Searle. New York: Columbia University Press.

Rubin, Miri. 1991. *Corpus Christi: The eucharist in late medieval culture*. Cambridge: Cambridge University Press.

Rudrauf, D., A. Lutz, D. Cosmelli, J. P. Lachaux, and M. Le Van Quyen. 2003. From autopoiesis to neurophenomenology: Francisco Varela's exploration of the biophysics of being. *Biological Research* 36 (1):27–65.

Saler, Benson. 1993/2000. *Conceptualizing religion: Immanent anthropologists, transcendent natives, and unbounded categories*. New York: Berghahn.

———. 2004. Towards a realistic and relevant "science of religion." *Method and Theory in the Study of Religion* 16 (3):205–33.

Satlow, M. L. 2005. Disappearing categories: Using categories in the study of religion. *Method and Theory in the Study of Religion* 17 (4):287–98.

Schachter, S., and J. Singer. 1962. Cognitive, social and physiological determinants of emotional state. *Psychological Review* 69:373–99.

Schacter, Daniel L. 2001. *The seven sins of memory: How the mind forgets and remembers*. Boston: Houghton Mifflin.

Schacter, Daniel L., and J. T. Coyle, eds. 1995. *Memory distortion: How minds, brains, and societies reconstruct the past*. Cambridge, MA: Harvard University Press.

Schmidt, Leigh Eric. 2000. *Hearing things: Religion, illusion, and the American enlightenment*. Cambridge, MA: Harvard University Press.

———. 2003. The making of modern "mysticism." *Journal of the American Academy of Religion* 71 (2):273–302.

Schneider, Susan. 2007. Daniel Dennett on the nature of consciousness. In *The Blackwell companion to consciousness*, edited by M. Velmans and S. Schneider. Malden, MA: Blackwell.

Schooler, Jonathan. 2002. Re-presenting consciousness: Dissociations between experience and meta-consciousness. *Trends in Cognitive Sciences* 6:339–44.

Searle, John. 2007. Putting consciousness back in the brain: Reply to Bennett and Hacker, *Philosophical foundations of neuroscience*. In *Neuroscience and philosophy*, edited by M. Bennett, D. C. Dennett, P. Hacker, and J. Searle. New York: Columbia University Press.

Segal, Robert A. 1983. In defense of reductionism. *Journal of the American Academy of Religion* 51 (1):97–124.

Shanafelt, Robert. 2004. Magic, miracle, and marvels in anthropology. *Ethnos* 69 (3):317–40.

Sharf, Robert H. 1995. Buddhist Modernism and the rhetoric of meditative experience. *Numen* 42 (3):228–83.

———. 1998. Experience. In *Critical terms for religious studies*, edited by M. C. Taylor. Chicago: University of Chicago Press.

———. 2000. The rhetoric of experience and the study of religion. *Journal of Consciousness Studies* 7 (11–12):267–87.

———. 2005. Ritual. In *Critical terms for the study of Buddhism*, edited by D. S. Lopez. Chicago: University of Chicago Press.

Sharpe, Eric J. 1986. *Comparative religion: A history*. 2nd ed. La Salle, IL: Open Court.

Shear, Jonathan. 2007. Eastern methods for investigating mind and consciousness. In *The Blackwell companion to consciousness*, edited by M. Velmans and S. Schneider. Malden, MA: Blackwell.

Sherwood, Simon J. 2002. Relationship between the hypnagogic/hypnopompic states and reports of anomalous experiences. *Journal of Parapsychology* 66: 127–50.

Shiota, Michelle, Dacher Keltner, and Amanda Mossman. 2007. The nature of awe: Elicitors, appraisals, and effects on self-concept. *Cognition and Emotion* 21 (5):944–63.

Shrady, Nicholas. 2008. *The last day: Wrath, ruin, and reason in the great Lisbon earthquake of 1755*. New York: Viking.

Steadman, Lyle B., Craig T. Palmer, and Christopher F. Tilley. 1996. The universality of ancestor worship. *Ethnology* 35 (1): 63–76.

Sidgwick, Eleanor. 1915. A contribution to the study of the psychology of Mrs. Piper's trance phenomena. *Proceedings of the Society for Psychical Research* 28:i–652.

Silberman, Israela. 2005. Religion as a meaning system: Implications for the new millennium. *Journal of Social Issues* 61 (4):641–63.

Slingerland, Edward G. 2008. *What science offers the humanities: Integrating body and culture*. Cambridge: Cambridge University Press.

Slone, D. Jason, ed. 2006. *Religion and cognition: A reader*. London: Equinox.

Smith, Jonathan Z. 1990. *Drudgery divine: On the comparison of early Christianities and the religions of late antiquity*. Chicago: University of Chicago Press.

———. 1998. Religion, religions, religious. In *Critical terms for religious studies*, edited by M. C. Taylor. Chicago: University Chicago Press.

———. 2000. Classification. In *A guide to the study of religion*, edited by W. Braun and R. McCutcheon. London and New York: Cassell.

Solms, Mark. 2000. Dreaming and REM sleep are controlled by different brain mechanisms. *Behavioral and Brain Sciences*. Special Issue: *Sleep and dreaming* 23 (6):843–50; 904–1018; 1083–1121.

Solomonova, E., T. Nielsen, P. Stenstrom, V. Simard, E. Frantova, and D. Donderi. 2008. Sensed presence as a correlate of sleep paralysis distress, social anxiety, and waking state social imagery. *Consciousness and Cognition* 17 (1):49–63.

Sørensen, Jesper. 2007. *A cognitive theory of magic*. Lanham, MD: Rowan and Littlefield.

Sosis, R., and C. Alcorta. 2003. Signaling, solidarity, and the sacred: The evolution of religious behavior. *Evolutionary Anthropology* 12 (6):264–74.

Sperber, Dan. 1996. Why are perfect animals, hybrids, and monsters food for symbolic thought? *Method and Theory in the Study of Religion* 8 (2):143–69.

Spilka, Bernard. 2003. *The psychology of religion: An empirical approach*. 3rd ed. New York: Guilford.

Spilka, Bernard, and Daniel N. McIntosh. 1995. Attribution theory and religious experience. In *Handbook of religious experience*, edited by J.R.W. Hood. Birmingham, AL: Religious Education Press.

———, eds. 1997. *The psychology of religion: Theoretical approaches*. Boulder, CO: Westview/Harper.

Spilka, Bernard, Phillip Shaver, and Lee A. Kirkpatrick. 1985. A general attribution theory for the psychology of religion. *Journal for the Scientific Study of Religion* 24 (1):1–20.

Spiro, Melford E. 1966. Religion: Problems of definition and explanation. In *Anthropological Approaches to the Study of Religion*, edited by M. Banton. New York: Tavistock.

Stark, Rodney. 1999a. Micro foundations of religion: A revised theory. *Sociological Theory* 17 (3):264–89.

———. 1999b. A theory of revelations. *Journal for the Scientific Study of Religion* 38 (2):287–308.

Stephen, Michele. 1989. The self, the sacred other, and the religious imagination. In *The religious imagination in New Guinea*, edited by G. Herdt and M. Stephen. New Brunswick, NJ: Rutgers University Press.

Stickgold, R., E. Paceschott, and J. A. Hobson. 1994. A new paradigm for dream research: Mentation reports following spontaneous arousal from REM and NREM sleep recorded in a home setting. *Consciousness and Cognition* 3 (1): 16–29.

Sun, Ron. 2006. *Cognition and multi-agent interaction: From cognitive modeling to social simulation*. Cambridge: Cambridge University Press.

Tajfel, Henri. 1981. *Human groups and social categories: Studies in social psychology*. Cambridge: Cambridge University Press.

Tanner, Amy. 1994. *Studies in spiritism*. Buffalo: Prometheus.

Taves, Ann. 1999. *Fits, trances, and visions: Experiencing religion and explaining experience from Wesley to James*. Princeton: Princeton University Press.

———. 2005. Religious experience. In *Encyclopedia of religion*, 2nd ed. Detroit: Thompson-Gale.

———. 2009a. William James revisited: Rereading *The Varieties of Religious Experience* in transatlantic perspective. *Zygon* 44 (2): 415–32.

———. 2009b. Channeled apparitions: On visions that morph and categories that slip. *Visual Resources* 25 (1): 141–56.

Tetlock, P. E. 2003. Thinking the unthinkable: Sacred values and taboo cognitions. *Trends in Cognitive Science* 7 (7):320–24.

Tetlock, P. E., O. V. Kristel, S. B. Elson, M. C. Green, and J. S. Lerner. 2000. The psychology of the unthinkable: Taboo trade-offs, forbidden base rates, and heretical counterfactuals. *Journal of Personal and Social Psychology* 78 (5):853–70.

Thomas, Keith. 1971. *Religion and the decline of magic*. London: Weidenfeld and Nicolson.

Thompson, Evan. 2001. *Between ourselves: Second-person issues in the study of consciousness*. Charlottesville, VA: Imprint Academic.

Thompson, Evan, and Francisco Varela. 2001. Radical embodiment: Neural dynamics and consciousness. *Trends in Cognitive Science* 5:418–25.

Tomasello, Michael. 1999. *The cultural origins of human cognition*. Cambridge, MA: Harvard University Press.

Tomasello, Michael, Malinda Carpenter, Josep Call, Tanya Behne, and Henrike Moll. 2005. Understanding and sharing intentions: The origins of cultural cognition. *Behavioral and Brain Sciences* 28:675–735.

Tononi, G. 2005. Consciousness, information integration, and the brain. In *Boundaries of consciousness: Neurobiology and neuropathology*, edited by Steven Laureys. Amsterdam: Elsevier Science.

Trevarthen, Colwyn, and Vasudevi Reddy. 2007. Consciousness in infants. In *The Blackwell companion to consciousness*, edited by Max Velmans and Susan Schneider. Malden, MA: Blackwell.

Tweed, Thomas A. 2006. *Crossing and dwelling: A theory of religion*. Cambridge, MA: Harvard University Press.

Twiss, Sumner B., and Walter H. Conser. 1992. *Experience of the sacred: Readings in the phenomenology of religion*. Hanover, NH: University Press of New England.

Tylor, Edward Burnett. 1970. *Religion in primitive culture.* Gloucester, MA: P. Smith.

Vail, Peter. 2004. Making the mundane sacred through technology: Mediating identity, ecology, and commodity fetishism. *Visual Communication* 3 (2):129–44.

Van Ness, Peter H. 1996. *Spirituality and the secular quest.* New York: Crossroad.

Varela, Francisco. 1995. The emergent self. In *The third culture: Beyond the scientific revolution,* edited by J. Brockman. New York: Simon and Schuster.

———. 1996. Neurophenomenology: A methodological remedy to the hard problem. *Journal of Consciousness Studies* 3:330–50.

Varela, Francisco, and Jonathan Shear, eds. 1999. *The view from within: First-person approaches to the study of consciousness.* Bowling Green, IN: Academic Imprints.

Varela, Franscisco J., Evan Thompson, and Eleanor Rosch. 1993. *The embodied mind: Cognitive science and human experience.* Cambridge, MA: The MIT Press.

Velmans, Max. 2001. Heterophenomenology versus critical phenomenology: A dialogue with Dan Dennett. Available online at http://.cogprints.soton.ac.uk/documents/disk0/00/00/17/95/index.htlm.

———. 2006. Heterophenomenology versus critical phenomenology. Available online at http://.cognprints.org./4741/.

Velmans, Max, and Susan Schneider, eds. 2007. *The Blackwell companion to consciousness.* Malden, MA: Blackwell.

Vygotsky, L. S. 1978. *Mind in society: The development of higher psychological processes.* Cambridge, MA: Harvard University Press.

Waghorne, Joanne. 1999. The divine image in contemporary India. In *Born in heaven, made on earth: The creation of the cult image in the ancient Near East,* edited by M. B. Dick. Winona Lake, IN: Eisenbrauns.

Wagner, Roy. 2005. Taboo. In *Encyclopedia of religion,* 2nd ed. Detroit: Thompson-Gale.

Wandel, Lee Palmer. 2006. *The eucharist in the Reformation: Incarnation and liturgy.* Cambridge, MA: Cambridge University Press.

Waterworth, J., ed. 1848. *The canons and decrees of the sacred and ecumenical Council of Trent.* London: Dolman.

Watt, Douglas. 2005. Attachment mechanisms and the bridging of science and religion. In *Ways of Knowing: Science and Mysticism Today,* edited by Chris Clarke. Charlottesville, VA: Academic Imprints.

———. 2007. Toward a neuroscience of empathy: Integrating affective and cognitive perspectives." *Neuro-Psychoanalysis* 9 (2):119–72.

Weiner, Annette. 1985. Inalienable wealth. *American Ethnologist* 12 (2):210–27.

Whitehouse, Harvey. 2004. *Modes of religiosity: A cognitive theory of religious transmission.* Walnut Creek, CA: AltaMira.

Wiebe, Donald. 1984. Beyond the sceptic and the devotee: Reductionism in the scientific study of religion. *Journal of the American Academy of Religion* 52 (1):157–65.

———. 1999. *The politics of religious studies: The continuing conflict with theology in the academy.* New York: Palgrave.

Wikström, Owe. 1987. Attribution, roles and religion: A theoretical analysis of Sundén's role theory of religion and the attributional approach to religious experience. *Journal for the Scientific Study of Religion* 26 (3):390–400.

Winnicott, Donald W. 1971. *Playing and reality*. London: Tavistock.

Wittgenstein, Ludwig. 1953. *Philosophical investigations*. New York: Macmillan.

Wittkower, Rudolf. 1942. Marvels of the East: A study in the history of monsters. *Journal of the Warburg and Courtald Institutes* 5:159–97.

Woody, Erik, and Henry Szechtman. 2000. Hypnotic hallucinations and yedasentience. *Contemporary Hypnosis* 17 (1):26–31.

Wright, Lawrence. 1994. *Remembering Satan: A tragic case of recovered memory*. New York: Vintage.

Zahavi, Dan. 2005. *Subjectivity and selfhood: Investigating the first-person perspective*. Cambridge, MA: The MIT Press.

Zaidman, Nurit. 2003. Commercialization of religious objects: A comparison between traditional and new age religions. *Social Compass* 50 (3):345–60.

Zelazo, Philip David, Helena Hong Gao, and Rebecca Todd. 2007. The development of consciousness. In *The Cambridge handbook of consciousness*, edited by Philip David Zelazo, Morris Moscovitch, and Evan Thompson. Cambridge: Cambridge University Press.

Zimbardo, Philip G. 1992. *Psychology and life*. 13th ed. New York: HarperCollins.

Zimdar-Swartz, Sandra L. 1991. *Encountering Mary: From La Salette to Medjugorje*. Princeton: Princeton University Press.

Zinnbauer, Brian J., and Kenneth I. Pargament. 2005. Religiousness and spirituality. In *Handbook of the psychology of religion and spirituality*, edited by R. F. Paloutzian and C. L. Park. New York: Guilford.

Name Index

Note: Concepts and other subjects can be found in the subject index. Page references set in italics refer to illustrations.

Subject Index

Note: Personal names can be found in the name index. Page references set in italics refer to illustrations.

absolutes, 36–38, 45. *See also* ideal things
activation-input-modulation (AIM) model, 76, 78, 164
adepts, 154–55. *See also* mediums
agency: ambiguity-/threat-inspired attributions of, 137–38; anomalous, 40–45, 45, 138, 178; detection, 135, 138; divine vs. human vs. demonic, 152–53. *See also* felt presence
altered states of consciousness, 21, 60, 82, 84–85, 110, 136, 164–65
American Psychiatric Association, 77
ancestor veneration, 44
anomalous things: 36, 38–39, 45, 54; with agency, 40–44 (*see also* agency); without agency, 39–40 (*see also* the mystical; the spiritual)
anthropology, 6–7
Ascending the Hall Ceremony, 51–53, 84–85
ascriptions: defined, 9, 181; distinguished from attributions, 10–11, 13, 19, 89n; individual vs. group, 53–55, 53 (table); simple vs. composite, 9–10, 12–14, 46–48, 53–55, 53 (table), 181, of specialness, 26–28. *See also* attribution theory; attributions; formations
—simple, 162–64. *See also* special things
—composite, 164–65. *See also* special paths
ascriptive model: arguments against, 20–21; vs. sui generis model, 17–22, 18 (table), 94. *See also* sui generis model; deeming things religious
atheists, 117
attribution theory, 13, 88–94; defined, 10–11, 19, 181; and levels of attribution, 111–18, 113 (table); of religion, 94–111; Malle on, 100–102; Park and Paloutzian on, 98, 100; Spilka, Shaver, and Kirkpatrick on, 94–95, 169–71; and

meaning-belief systems (MBS), 94–97; and the perennialist-constructivist debates, 91–95, 92n, 116; and top-down vs. bottom-up processing, 94, 98–99, 109
attributions: defined, 9, 181; distinguished from ascriptions, 10–11, 13, 19, 89n. *See also* attribution theory
Augsburg, 143–44
authenticity, judging, 158
automatic behaviors, 60–61
awareness. *See* consciousness

beauty, 36–37, 178
beguines, 151–52, 151n
benevolence, eternal and free, 178
bliss, eternal, 178
bodies (objects) vs. persons, 44
Buddhism: Ascending the Hall Ceremony, 51–53, 84–85; Chan enlightenment, 51–53, 84–85; of the Dalai Lama, 37; *mārga*, 46, 81, 178–79; meditation practices of, 82, 84–85; religious experience as a source of authority for, 4; textual tradition of, 81; *vipassana* vs. *kensho*, 117n
building-block approach to the study of religion, 13–14, 161–65

Calvinism, 144, 146–47
Catholic Church: in Augsburg, 143–44; Benediction of the Blessed Sacrament, 147; Council of Trent, 144–46; Feast of Corpus Christi, 147, 151; hierarchy in, 115; on holy objects, 30–31; Lourdes (shrine), 30–32; mass deemed efficacious by, 143–44 (*see also* Eucharist, debates about); medieval visualization practices vs. neopaganism, 150, 153, 155–56; vs. Protestants, 144–47 (*see also* Eucharist, debates about); and ritual efficacy, 148; Society of Jesus, 155